NONLINEAR TIME SERIES ANALYSIS

Methods and Applications

T0332924

NONLINEAR TIME SERIES AND CHAOS

Editor: Howell Tong

Vol. 1: Dimension Estimation and Models
       ed. H. Tong

Vol. 2: Chaos & Forecasting
       ed. H. Tong

Forthcoming

Vol. 3: Nonlinear Time Series & Economic Fluctuations
       S. Potter

Nonlinear Time Series and Chaos Vol. 4

NONLINEAR TIME SERIES ANALYSIS

Methods and Applications

Cees Diks
University of Kent, England

World Scientific
Singapore • New Jersey • London • Hong Kong

Published by

World Scientific Publishing Co. Pte. Ltd.

P O Box 128, Farrer Road, Singapore 912805

USA office: Suite 1B, 1060 Main Street, River Edge, NJ 07661

UK office: 57 Shelton Street, Covent Garden, London WC2H 9HE

Library of Congress Cataloging-in-Publication Data
Diks, Cees.
    Nonlinear time series analysis : methods and applications / Cees
Diks.
        p.   cm. -- (Nonlinear time series and chaos : vol. 4)
    Includes bibliographical references and index.
    ISBN 9810235054 (alk. paper)
    1. Time-series analysis.   2. Nonlinear theories.   I. Title.
II. Series.
QA280.D55   1999
519.5'5--DC21   99-16722
                                              CIP

British Library Cataloguing-in-Publication Data
A catalogue record for this book is available from the British Library.

Printed in Singapore by Regal Press (S) Pte. Ltd.

FOREWORD

The present volume of *Nonlinear Time Series and Chaos* gives me the opportunity to present a number of methods for the analysis of time series. Although these methods were originally inspired by the need for tools for analyzing the short and noisy nonlinear time series one is often confronted with in physiology, they have since also found several applications in other fields.

The currently established methods for the analysis of time series were developed mainly in two fields: statistical time series analysis and the theory of dynamical systems. A prolific cross-fertilization has recently started to develop between these areas, and it is my intention to present some theory and methods in view of this development, indicating the connections between the two fields wherever possible. Statistics has a long tradition of studying joint probability distributions of time series. The analog of this approach in chaos theory is the study of reconstruction measures. As reconstruction measures can be defined both for deterministic time series and stochastic time series they provide a natural framework for a general approach.

Throughout, I have tried to keep the material accessible to a readership with an interest in nonlinear time series analysis from a wide variety of research areas; only modest mathematical background knowledge is assumed and examples are given throughout to illustrate the ideas behind the methods presented. After a brief introduction to the theory of nonlinear time series analysis, a test for reversibility and a test for comparing two time series are described. Then deterministic time series with observational noise are considered and methods are proposed for estimating the noise level, fractal dimension, and entropy of the dynamics simultaneously. Finally two spatio-temporal dynamical systems are analyzed.

I am indebted to Professor H. Tong for offering me the opportunity to write this book and for his encouragement and advice. I thankfully acknowledge Elsevier Science for permission to reproduce Figs 9.2, 9.6 and 9.7 and the American Physical Association for permission to reproduce Figs 6.2, 6.3 and Table 6.1. The work involved research which was partially supported by the Engineering and Physical Sciences Research Council. I gratefully acknowledge Professor W. R. van Zwet, Professor F. Takens and Dr J. DeGoede for their enthusiasm and help in developing much of the material that served as the basis for this book and Qiwei Yao, Catriona Queen and Catarina Resende for their suggestions for improving the manuscript.

Canterbury, *Fall 1998*   Cees Diks

CONTENTS

Foreword   v

1 Introduction   1

2 Nonlinear Dynamical Systems   7

3 Stochastic Time Series   29

4 A Test for Reversibility   51

5 Detecting Differences between Reconstruction Measures   81

6 Estimating Invariants of Noisy Attractors   97

7 The Correlation Integral of Noisy Attractors   117

8 Spiral Wave Tip Dynamics   127

9 Spatio-temporal Chaos: a Solvable Model   149

Appendix A   175

Appendix B   181

Appendix C   187

References   191

Index   207

CHAPTER 1

INTRODUCTION

Many phenomena in our environment are studied using sequences of measurements, or observations, made in the course of time. These sequences of observations, called time series, often comprise an important part of the information available on a system of under study. The analysis of time series is of relevance to a broad range of research areas as indicated by the variety of time series studied in the sciences. Examples include daily temperature records, electrocardiograms and exchange rates, among many others.

Observed time series often exhibit irregular and apparently unpredictable behavior. The main aims of time series modeling are to capture the essential features of the observed irregularity and to increase our understanding of the generating processes, or dynamics, of observed time series. The modeling of time series by linear stochastic processes, well known to statisticians, has a long history, and by now there is a vast literature on parameter estimation, optimal prediction and control for linear stochastic systems. Linear Gaussian stochastic time series models in particular have received much attention, mainly because of their convenient analytic properties.

In contrast, the evolution of deterministic dynamical systems does not depend on a noise source. For those systems, there is a deterministic way, often modeled by difference equations or by differential equations, in which future states are determined by the present state. By now it is well established that simple deterministic systems with only one or a few state variables may already give rise to rich and complex behavior. This phenomenon, now referred to as chaos, was known already to Poincaré at the turn of the twentieth century, and did not receive much attention in science until a few decades ago. The popularity of chaos is at least partially caused by the availability of fast and cheap computer power which has enabled research by experimental mathematics. Chaos theory has produced a wealth of powerful methods for the analysis of deterministic time series, including methods for the reconstruction of the dynamics from an observed time series and for estimation of characteristics such as the fractal dimension of an attractor of the dynamics. Concurrently

1

with the development of chaos theory, the role of nonlinearity has also become widely recognized among statisticians, as many properties observed in experimental time series, such as time irreversibility, simply can not be explained by linear Gaussian models. The importance of nonlinearity in statistical modeling is emphasized notably by Tong (1990).

Clearly, both linear time series analysis and chaos analysis are of limited use when applied to the short, nonlinear, and noisy time series as usually encountered in real experiments. In many fields, such as the life sciences, but also the chemical and physical sciences, long and noise-free time series are often beyond experimental reach. Our motivation is the study of such nonlinear experimental time series within a framework that is based on concepts from both chaos theory and traditional statistical time series analysis. The idea is that concepts from chaos theory can improve our understanding of the types of behavior observed in nonlinear stochastic systems, whereas the theory of statistics can provide insights into the sampling properties of estimators, and motivate new estimators for the characterization of time series.

1. Notation

The set of reals is denoted by $\mathbf{R}$, while the m-dimensional Euclidean space is denoted by $\mathbf{R}^m$. Vectors in $\mathbf{R}^m$ will be denoted by bold face symbols such as x, that is, $x = (x_1, x_2, \ldots, x_m)^T$ where T denotes the transpose, while uppercase symbols like X refer to random variables. A time series consisting of N consecutive observations X_n is denoted by

$$\{X_n\}_{n=1}^N,$$

and we will often omit the range of the index when it is clear from the context; the observations in a time series of length N by default will have indices running from 1 up to and including N. The use of capital letters for the consecutive values in a time series indicates that, formally, we consider them as random variables. Time series that are entirely deterministic, with a known initial state, easily fit in as special cases for which each of the observations can take only one predetermined value. Occasionally, when we want to emphasize the deterministic nature of a time series, we will use lower case letters, as in $\{x_n\}$.

We will consider only time series with a constant time interval Δt between successive measurements so that the time t_n, at which measurement n is performed, is given by $t_n = t_0 + n\Delta t$. Throughout, the observations are assumed to be real-valued scalar quantities. Generalizations to multivariate time series usually are relatively straightforward and will not be discussed explicitly.

2. About Reading this Book

The following chapters of this book consist of two introductory chapters and six chapters in which a number of recently developed methods in the field of time series analysis are described, applied and discussed. As most of the chapters evolved from separate practical problems they can be read independently without difficulty. This, however, does neither imply that cross-references between different chapters are absent nor that they are irrelevant, as adopted approaches are frequently motivated by results and discussions presented in other chapters.

2.1. Introductory Chapters

Chapter 2 presents a brief informal overview of the theory of dynamical systems and deterministic time series. Guided by examples, it introduces the notions of chaos and strange attractors, and discusses the reconstruction of attractors from an observed time series and estimation of the correlation dimension and the correlation entropy with the Grassberger-Procaccia method.

Chapter 3 describes time series with observational noise, dynamical noise and time series generated by spatio-temporal dynamical systems. The (spurious) results that one may obtain when applying chaos analysis to these non-deterministic time series are discussed. The concepts of reconstruction measures and prediction measures are looked at from the dynamical systems point of view as well as the statistical point of view. This provides a perspective for the remaining chapters, in which a number of specific practical problems are studied extensively and some new methodology is developed.

2.2. Subsequent Chapters

The desirability to construct methods for the nonlinear analysis of time series that are not explicitly based on the assumptions of linearity or determinism has led to the three methods described in Chapters 4, 5, and 6. These methods by no means are intended to provide a full characterization of time series but may be considered as tools for analyzing time series, motivated both by dynamical systems theory and statistics, and requiring little assumptions concerning the nature of the generating mechanism of the time series.

Chapter 4 describes a test for the hypothesis that an observed stationary time series is reversible. If irreversibility is detected, linear Gaussian random processes as well as static transformations of linear Gaussian random processes can be excluded as the generating mechanism of the time series. After introducing a test statistic based on a distance notion between time forward and time backward reconstruction measures, a block method is introduced for taking into account dependence among reconstruction vectors in the calculation of the variance of the test statistic.

In Chapter 5 a test is described for the hypothesis that two stationary time series are independent realizations of identical stationary random processes. The test is based on the same distance notion between reconstruction measures as used in the test for reversibility. Also here the dependence among reconstruction vectors can be handled using a block method.

Chapter 6 presents a method for the analysis of low-dimensional deterministic time series in the presence of independent, identically distributed (i.i.d.) Gaussian observational noise. It is based on a Gaussian kernel correlation integral, from which the correlation dimension, the correlation entropy and the noise level can be estimated simultaneously using a nonlinear fit procedure. Applications to computer generated time series indicate that the method works well for noise levels of up to 20% (in standard deviation). For some empirical time series, the model can be fitted well whereas for others the poor fits suggest that they are not described well as low-dimensional time series with observational noise.

In Chapter 7 the problem of observational noise is examined again but from a slightly different angle. An expression is derived for the usual correlation integral based on the analytic expression for the Gaussian kernel correlation integral found in Chapter 6. The results are consistent with those of Smith (1992b) and Oltmans and Verheijen (1997) who derived expressions for the correlation integral in the presence of Gaussian observational noise using different approaches.

The last two chapters explore the limits of traditional chaos analysis using time series generated by spatio-temporal models. Chapter 8 analyzes tip trajectories of meandering and hyper-meandering spiral waves in Barkley's model: a two-dimensional excitable media model, governed by a two-variable partial differential equation, which allows rotating spiral wave excitations as solutions. It is examined whether the tip position time series of these spiral wave excitations are low-dimensional deterministic, and possibly chaotic, as suggested by Jahnke and Winfree (1991). Two tip position time series, generated for different choices of the model parameters, and one of which looks regular while the other looks irregular, are analyzed with the Grassberger-Procaccia method. For both parameter values the behavior of the tip velocity time series turns out to be simpler than the tip position time series. This can be explained by a simple model that reproduces many qualitative features of the observed tip trajectories.

Chapter 9 systematically studies a model exhibiting spatio-temporal chaos. The piece-wise linearity of the model enables analytic calculation of the spectrum of Lyapunov exponents and the construction of a phase diagram showing the regions in parameter space for which the model is spatio-temporally chaotic. For several parameter values the time series obtained from a single site have identical reconstruction measures, even though the information entropy densities are different. It follows that there are systems (albeit possibly of a non-generic type) which can not be fully characterized by methods based on the analysis of time series measured at a single site only, as suggested in some recent publications.

CHAPTER 2

NONLINEAR DYNAMICAL SYSTEMS

This chapter lays out some concepts and ideas basic to the analysis of deterministic time series, often referred to as chaos analysis. Many of the recently developed methods for the analysis of deterministic time series are based on results in the theory of dynamical systems. Some terminology is introduced while discussing a number of examples of dynamical systems. Subsequently, the reconstruction theorem of Takens and the estimation of dynamical invariants are described.

1. Dynamical Systems

Before giving a formal definition of a dynamical system an example of a simple discrete time dynamical system is presented.

1.1. Example: The Logistic Map

The logistic map, given by

$$x_{n+1} = ax_n(1 - x_n), \tag{2.1}$$

may be interpreted as a simple biological model for the evolution of the population size x of some species from generation to generation. The population size of the next generation depends on the difference between two terms, one of which is linear, and the other quadratic in the previous population size. The linear term describes growth (at a rate $a > 1$) of the population size in the presence of unlimited vital resources, whereas the quadratic term accounts for the competition for vital resources among individuals. Because resources are limited there is a maximum population size which in suitable units is equal to 1. The population size is nonnegative, so that the state space Ω for this model may be taken to be the interval $[0, 1]$. To restrict the population size to this interval, a is assumed to be no larger than 4.

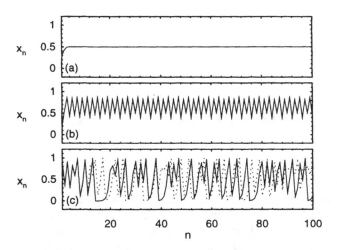

Figure 2.1: *Time series generated with the logistic map for parameter values $a = 2$ (a), $a = 3.5$ (b) and $a = 4$ (c). The initial conditions for the evolutions shown in (c) were $x_0 = 0.9$ (solid line) and $x_0 = 0.9001$ (dotted line).*

Figure 2.1 shows time series consisting of consecutive x-values of the logistic map for several values of the parameter a. We are interested in the asymptotic (long-term) behavior of the model rather than in its transient initial behavior. Clearly, the asymptotic behavior observed for the logistic map varies considerably with the parameter value a; for $a = 2$ the asymptotic evolution is confined to the stable *fixed point* solution $x = 0.5$, for $a = 3.5$ it is periodic with a period of 4 time units while the typical evolutions for $a = 4$ have an erratic nature. The latter is an example of *chaotic* behavior, in which the evolution depends in a sensitive manner on the initial state. Two evolutions starting from slightly different initial states are very similar for the first few iterations, but the difference between the two states grows exponentially and reaches macroscopic scales within a finite number of iterations. This is illustrated in Figure 2.1 (c), which shows the evolutions obtained for two initial conditions that differ only by 10^{-4}.

Formally, closed-form expressions for the evolutions of the logistic map can be found by solving the so-called Poincaré equation associated with the evolution rule given in Equation (2.1):

$$F(cz) = aF(z)(1 - F(z)). \tag{2.2}$$

The solution of Equation (2.1) then becomes $x_n = F(c^n\theta)$ where θ must be taken in accordance with the initial state x_0 through $F(\theta) = x_0$.

The solutions of the Poincaré equation of the logistic map for general values of a are unknown, but for some specific values of a a solution can be easily found. For $a = 2$, a solution of the Poincaré equation is $F(z) = \frac{1}{2}(1 - e^{-z})$, as it satisfies $F(cz) = 2F(z)(1 - F(z))$ for $c = 2$. We obtain the evolutions

$$x_n = \frac{1}{2}(1 - e^{-2^n\theta}), \qquad (2.3)$$

describing fast convergence to the stable fixed point $x = \frac{1}{2}$. For $a = 4$ (a chaotic case) a solution of the Poincaré equation can be found from the doubling rule of the sine function, which reads $\sin(2z) = 2\sin(z)\cos(z)$. Squaring this relation gives $\sin^2(2z) = 4\sin^2(z)\cos^2(z) = 4\sin^2(z)(1 - \sin^2(z))$. Comparing this with Equation (2.2), one can see that the choice $c = 2$ provides a solution $F(z) = \sin^2(z)$ for the case $a = 4$. Therefore,

$$x_n = \sin^2(2^n\theta) \qquad (2.4)$$

is an analytic expression for the evolution of the logistic map, which is consistent with the initial condition if θ is taken to be one of the solutions of $\sin^2(\theta) = x_0$. This result shows that a chaotic dynamical system is not inconsistent with an analytic expression for the future states.

From a population dynamical point of view, the logistic map is of course extremely simple. It is an idealized nonlinear model for the dynamics of a single population size only. The example is given merely to illustrate that already a very simple evolution rule involving only one state variable can give rise to irregular evolutions. For other population dynamical models, including multiple species models such as predator-prey models, we refer to the book by Murray (1989).

Definition of a Dynamical System   A *dynamical system* consists of a triple (Ω, ϕ, T), where Ω is a *state space* or *phase space* representing all possible states of the system, so that at each time $t \in T$, where T denotes the set of possible times, the system is in one of the possible states $x(t) \in \Omega$. T has the semigroup property, that is, for $s, t \in T$, also $s + t \in T$. A dynamical system furthermore has an *evolution operator*,

$$\phi : \Omega \times T \to \Omega. \qquad (2.5)$$

For each fixed t, $\phi(x, t)$ defines a map

$$\phi^t : \Omega \to \Omega, \qquad \phi^t(x) = \phi(x, t), \qquad (2.6)$$

called the *flow* over time t. For fixed $x \in \Omega$, $\phi(x, t)$ is called the *evolution* of x or the *trajectory* through x.

For *invertible* dynamical systems the flow has the group properties

$$\phi^0 = I, \tag{2.7}$$

and

$$\phi^s \circ \phi^t = \phi^{s+t}, \tag{2.8}$$

for arbitrary $s, t \in T$. If the dynamical system is non-invertible, these relations only hold for positive values of s and t, and the flow is said to have semi-group properties. □

The set T of allowed times may be either discrete, e.g. $T \sim \mathbf{Z}$, or continuous, e.g. $T \sim \mathbf{R}$. In the former case we speak of a discrete time dynamical system, and in the latter of a continuous time dynamical system. Discrete time dynamical systems are often defined in terms of a map $f = \phi^1$, which represents the flow over a time unit interval, and evolutions can be obtained by successive application, or iteration, of this map. Continuous time dynamical systems are often defined in terms of a set of ordinary differential equations, and the evolutions of these systems consist of the integrals of the equations through the initial state x_0.

1.2. Example: The Doubling Map

From the representation in Equation (2.4), it immediately follows that the values x_n of the logistic map with $a = 4$ can be written alternatively as $x_n = h(y_n)$ where y_n is given by

$$y_{n+1} = 2y_n \bmod 1, \qquad \text{with } y_0 = \theta/\pi \tag{2.9}$$

by choosing the *read-out function* or *measurement function* $h(y)$ to be $h(y) = \sin^2(\pi y)$. Alternatively, it can be readily checked that the analytic solutions $y_n = 2^n\theta/\pi \bmod 1$ of Equation (2.9) give $x_n = h(y_n) = \sin^2(2^n\theta)$ in accordance with Equation (2.4). The model given in Equation (2.9) is known as the doubling map.

The doubling map has a nice interpretation when the state variables y_n are expanded in binary notation (i.e. in base 2), giving

$$y_n = 0.s_1 s_2 s_3 \ldots, \tag{2.10}$$

say, where each of the digits s_i is either 0 or 1. Using this binary expansion one iteration can be described as follows. The multiplication by a factor of two shifts every bit in the expansion of y_n to the left by one position. The modulo operation then discards the first digit preserving the fractional part of the expansion, giving

$$y_{n+1} = 0.s_2 s_3 s_4 \ldots, \tag{2.11}$$

which nicely demonstrates the notion of *sensitive dependence on initial conditions*. Starting from two slightly different initial conditions, the binary expansions of which match up to the kth digit, say, the trajectories will be very close during the first few iterations. In subsequent iterations, the first digit for which a difference in the expansions occurs becomes more and more significant, until, after k iterations, the difference between the two evolutions has become of the order of the range of the time series. This shows why, in order to perform reliable predictions over arbitrary long periods of time, the initial state must be known with infinite precision. Because the binary expansions of almost all numbers between 0 and 1 have non-periodic tails, the evolution from a random initial point y_0, where y_0 is picked randomly from $U[0,1]$, is asymptotically non-periodic with probability one.

1.3. Example: The Hénon Map

A generalization of the logistic map to a two-dimensional dynamical system is the Hénon map (Hénon, 1976), given by

$$\begin{aligned} x_{n+1} &= 1 - ax_n^2 + y_n \\ y_{n+1} &= bx_n. \end{aligned} \tag{2.12}$$

The Jacobian of the mapping is

$$J = \begin{pmatrix} -2ax_n & 1 \\ b & 0 \end{pmatrix}, \tag{2.13}$$

which has determinant $-b$, which implies that the map is state space volume contracting (termed *dissipative* in the physics literature) for $|b| < 1$. Here, the parameter values $a = 1.4$ and $b = 0.3$ will be used, corresponding to the case most studied in the literature. As the map is state space volume contracting and evolutions are bounded within a certain region around the origin (Ruelle, 1980), all evolutions starting from within this region asymptotically are confined to a set with zero volume, called the *attractor* of the dynamical system. Figure 2.2 shows the attractor of the Hénon map. The figure consists of the points (x_n, y_n) for 4000 consecutive iterations of the map starting from $(0,0)$, after discarding the first 100 iterations to allow for the evolution to settle

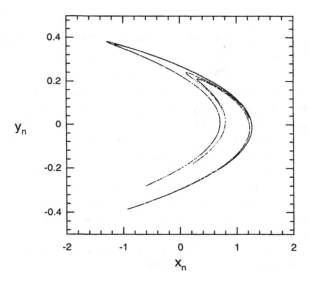

Figure 2.2: *Attractor of the Hénon map.*

onto the attractor. The attractor shows a nontrivial complex structure, and such attractors (that consist of neither a finite number of points, nor of an n-torus) are often called *strange attractors*, a term coined by Ruelle and Takens (1971). Although the Hénon map is often used as an example of a simple dynamical system with a strange attractor, it was only recently proved that it has a strange attractor for a set of parameters (a, b) of positive Lebesgue measure (volume) (Benedicks, 1994). Whether the Hénon map with standard parameter values has a strange attractor rather than a periodic attractor with a very long period is still an open problem.

1.4. Example: The Lorenz System

The three-variable model proposed by Lorenz (1963) is one of the first chaotic dynamical systems discovered. It is a continuous time dynamical sys-

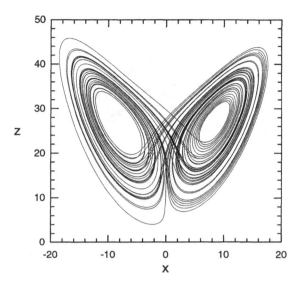

Figure 2.3: *Projection on the x-z plane of a trajectory of the Lorenz dynamical system with $b = 8/3$, $\sigma = 10$ and $r = 28$.*

tem given by the set of ordinary differential equations

$$
\begin{aligned}
\frac{\mathrm{d}x}{\mathrm{d}t} &= \sigma(y - x) \\[2mm]
\frac{\mathrm{d}y}{\mathrm{d}t} &= (r - z)x - y \\[2mm]
\frac{\mathrm{d}z}{\mathrm{d}t} &= xy - bz,
\end{aligned}
\tag{2.14}
$$

where $b = 8/3$, σ is the Prandtl number, and r is an external control parameter which is proportional to the Rayleigh number. The equations represent a simplified three-variable model of a forced dissipative hydrodynamic flow (Rayleigh-Bénard convection) in which x represents the flow speed, and y and z describe the temperature profile. A short derivation of the Lorenz equations can be found in the book by Schuster (1988). A typical projection of a trajectory on the x-z plane is shown in Figure 2.3 for $\sigma = 10$ and $r = 28$.

The Lorenz system has three *unstable fixed points*: the origin, $(x, y, z) = (0, 0, 0)$ and the symmetric pair $(x, y, z) = (\pm\sqrt{b(r-1)}, \pm\sqrt{b(r-1)}, r-1) \approx (\pm 8.5, \pm 8.5, 27.0)$. The projections on the x-z plane of the latter two fixed

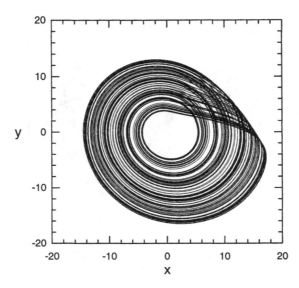

Figure 2.4: *Projection on the x-y plane of a trajectory of the Rössler dynamical system.*

points lie in the centers of the two wings of the attractor shown in Figure 2.3. The evolution keeps moving around this pair of fixed points in an apparently irregular manner; the evolution spirals outwards around one of the two unstable fixed points until the state reaches the vicinity of the other unstable fixed point. After that the evolution starts to spiral outwards around the other fixed point, and the scenario is repeated with the roles of the two fixed points reversed. The evolution is chaotic, exhibiting sensitive dependence on initial conditions and never exactly repeating previously followed paths.

1.5. Example: The Rössler System

The Rössler model (1976) was designed to capture the essential chaotic characteristics of the Lorenz model and is sometimes used as a caricature for the macroscopic equations of a stirred chemical reaction sustained by a flow of chemicals, without which the system would approach a stable equilibrium state (Van Kampen, 1992). The model equations are

$$\frac{dx}{dt} = -y - z$$

$$\frac{dy}{dt} = x + ay \qquad (2.15)$$

$$\frac{dz}{dt} = b + z(x - c).$$

Figure 2.4 shows the projection on the x-y plane of the Rössler attractor for the standard parameter values $a = 0.15$, $b = 0.2$ and $c = 10$. Most of the time, the evolution takes place close to the x-y plane. However, the equations frequently gives rise to large outbursts in the z-direction, after which the evolution is injected near the x-y plane again. In fact, a stretching and folding (horseshoe map) is responsible for the chaotic behavior of the Rössler system.

2. The Reconstruction Theorem

The reconstruction method (Packard *et al.*, 1980; Takens, 1981) enables the reconstruction of the asymptotic dynamics of a dynamical system from an observed deterministic time series. Many of the tools used for analysis of deterministic time series rely on this method in one way or another. Here we will give an informal introduction to the reconstruction method without discussing the proof of the *reconstruction theorem* (Takens, 1981) in detail; the line of thought is indicated and illustrated with some examples.

The starting point for the reconstruction theorem is a dynamical system described by an unknown differential equation

$$\frac{dy}{dt} = F(y), \qquad y \in \Omega, \qquad (2.16)$$

with state space Ω. In the proof of the reconstruction theorem it is assumed that the evolutions of the system are confined to a finite-dimensional bounded subspace M of the (possibly infinite dimensional) phase space Ω. The measurement function h is assumed to be smooth and to depend on the state of the dynamical system only, that is, it depends on time only implicitly through the time dependence of the state variable. For simplicity, we will assume that h is a real-valued scalar function, that is, h is a map from M to the real line **R**.

Firstly, we note that, for each n, observation x_n is the image of the state $y_n \in M$ under the map $h : M \to \mathbf{R}$ defined by $y \mapsto h(y)$. Secondly, the state

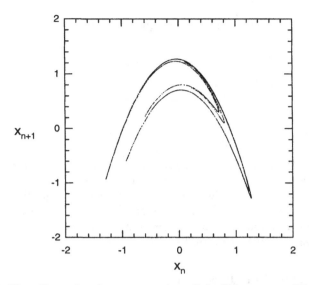

Figure 2.5: *Two-dimensional reconstruction of the Hénon map. There is a one-to-one correspondence between the reconstructed attractor and the original attractor shown in Figure 2.2*

$y_n \in M$ uniquely determines the future states of the dynamical system, so that

$$\Phi_2 : M \to \mathbf{R}^2, \qquad y \mapsto (h(y), h(F(y))) \tag{2.17}$$

is a map from M to the two-dimensional plane, $\mathbf{R}^2$, which takes the state y_n to (x_n, x_{n+1}). An example of such a two-dimensional projection is shown in Figure 2.5 for the Hénon map. It shows a structure similar to that observed in the original phase space. The map Φ_2 for the Hénon map in fact is smooth and invertible, so that there is a one-to-one correspondence between the original states y and their images $\Phi_2(y)$. The attractor is mapped in a one-to-one fashion to its image called the *reconstructed attractor*.

A reconstruction provides one not only with a one-to-one image of the original attractor, but also with a representation of the dynamics, called the *reconstructed dynamics*. For the present example, we can derive explicit equations for the reconstructed dynamics. Since two consecutive values x_{n-1} and x_n uniquely determine the state of the Hénon system, the dynamics given in Equation (2.12) can be expressed in terms of the x-values only. Upon doing so, one obtains the single equation

$$x_{n+1} = 1 + bx_{n-1} - ax_n^2, \qquad (2.18)$$

expressing future values of x in terms of present and past values of x. This clearly demonstrates that the two-dimensional projection defined by Equation (2.17) is a complete representation of the original dynamics.

Often one needs to go to higher dimensional maps to obtain a reconstruction of a dynamical system. One can continue the line of thought displayed above by defining m-dimensional maps $\Phi_m : M \to \mathbf{R}^m$ through

$$\Phi_m : y \mapsto (h(y), h(F(y)), \ldots, h(F^{m-1}(y))), \qquad (2.19)$$

which maps y_n to $(x_n, \ldots, x_{n+m-1})$. The reconstruction theorem states that for smooth measurement functions h and for m sufficiently large, Φ_m generically is a smooth invertible map from M to $\mathbf{R}^m$ with a smooth inverse. It follows that the points represented by the *delay vectors* $(x_n, \ldots, x_{n+m-1})$ lie on a faithful image of the attractor in $\mathbf{R}^m$, referred to as the reconstructed attractor. Takens proved that it is sufficient to have $m \geq 2D + 1$ where $D = \dim(M)$ is the dimension of M. Several other inequalities can be derived involving the embedding dimension, the details of which we will not delve into here. We refer to Takens (1981) for details on the proof of the reconstruction theorem, and to Noakes (1991), Sauer *et al.* (1991), Hunt *et al.* (1992) and Sauer and Yorke (1993) for further discussions. Aeyels (1981, 1982) obtained some results related to the reconstruction by considering observability of nonlinear dynamical systems.

More generally, delay vectors $(x_n, x_{n+\tau}, \ldots, x_{n+(m-1)\tau})$ with arbitrary delay times τ can be used as these also give a faithful representation of the attractor generically. Of course, one can also construct delay vectors for non-deterministic time series, in which case the reconstruction theorem does not apply. By a slight abuse of language we will often refer to delay vectors as *reconstruction vectors*, even if the time series is not deterministic.

3. Correlation Dimension and Correlation Entropy

The reconstructed attractor has some geometrical structure, endowed with a measure related to the relative frequencies with which different parts of the attractor are visited. The measure associated with the attractor of a dynamical system is invariant under the evolution operator, and accordingly called the *invariant measure*.

The invariant measure can be characterized using various approaches. The reconstruction theorem shows that we can reconstruct the dynamics of a deterministic system up to a smooth parameter transformation, so that a natural approach for characterizing the dynamics would be one that is based on characteristics that are invariant under such transformations. Dimensions and entropies have become popular for characterizing attractors, and are invariant under the reconstruction procedure. Dimensions and entropies characterize the probability measures, μ_m associated with the reconstructed attractor at embedding dimension m of a stationary time series. The measure $\mu_m(A)$ of a given subset A of $\mathbf{R}^m$ is the relative asymptotic frequency with which reconstruction vectors visit A.

The correlation dimension and correlation entropy in particular are popular among time series analysts, as they can be estimated relatively straightforwardly from correlation integrals. The correlation integral $C_m(r)$ at embedding dimension m and radius r of a reconstruction measure μ_m is defined as the probability that the distance between two points drawn independently according to μ_m is smaller than r, that is,

$$C_m(r) = \iint \Theta(r - \|\boldsymbol{x} - \boldsymbol{y}\|) \, \mathrm{d}\mu_m(\boldsymbol{y}) \, \mathrm{d}\mu_m(\boldsymbol{x}), \qquad (2.20)$$

where $\Theta(\cdot)$ is the Heaviside function

$$\Theta(s) = \left\{ \begin{array}{ll} 0 & \text{for } s < 0 \\ 1 & \text{for } s \geq 0, \end{array} \right. \qquad (2.21)$$

and $\| \cdot \|$ is some norm, which is usually taken to be either the supremum norm $\|(x_1, \ldots, x_m)^T\| = \sup_i |x_i| = \max(|x_1|, \ldots, |x_m|)$, or the Euclidean norm $(\sum_{i=1}^m x_i^2)^{1/2}$.

For deterministic time series, the correlation integrals for small r and large m behave according to the scaling relation

$$C_m(r) \sim e^{-m\tau K_2} r^{D_2} \qquad (2.22)$$

where τ is the delay time used in the reconstruction, D_2 is the *correlation dimension* and K_2 is the correlation entropy per time unit, often abbreviated to *correlation entropy*. The correlation dimension D_2 is a dimensionless quantity which can be interpreted as the dimension of the attractor. It needs not be an integer and is bounded by the number of state variables of the system. It is a rough measure of the effective number of degrees of freedom or the effective number of variables involved in the generating process of the time series. The correlation entropy K_2, which has the dimension of one over time, is a measure of the rate at which pairs of nearby orbits diverge. It is often quoted in nats per time unit or bits per time unit depending on whether a base-e (natural) or a base-2 logarithm is used (we use the natural logarithm throughout). The quantity $1/K_2$ is a rough measure of the time scale on which errors increase by a factor e as a result of the dynamics. A large value of K_2 implies that, from a state known with finite precision, accurate predictions can be made only over a short number of time steps ahead and K_2 may thus be viewed as a measure of the unpredictability of a time series. In thermodynamics, the concept of entropy is also used.

The connection between the concepts of entropy used in the theory of dynamical systems and in thermodynamics is described by Petersen (1983) and by Schuster (1988). In both contexts, the entropy is an information theoretical quantity used to describe the amount of disorder, that is, the amount of information required to specify the state of a system up to a given accuracy. The entropy per time unit used in the theory of dynamical systems is a measure of the rate at which information about past states of dynamical systems is lost in the course of time.

3.1. The Grassberger-Procaccia Method

We next describe the estimation of D_2 and K_2 from an observed time series by the Grassberger-Procaccia method (Grassberger and Procaccia, 1983a; 1983b). From an observed time series $\{X_n\}_{n=1}^L$, of length L, sets of $N = L - (m-1)\tau$ reconstruction vectors, $\boldsymbol{X}_n = (X_n, X_{n+\tau}, \ldots, X_{n-(m-1)\tau})$ are constructed for consecutive embedding dimensions m. For each value of m, the set of delay vectors constitutes a cloud of points in $\mathbf{R}^m$, which for stationary time series can be regarded as a sample from the underlying m-dimensional probability measure μ_m. The probability measure which assigns equal mass $1/N$ to each of the observed delay vectors $\boldsymbol{X}_n$ is called the empirical delay vector measure, or empirical reconstruction measure.

In the Grassberger-Procaccia method the correlation integrals, defined in Equation (2.20), are estimated as

$$\widehat{C}_m(r) = \frac{2}{N(N-1)} \sum_{i=1}^{N-1} \sum_{j=i+1}^{N} \Theta(r - \|\boldsymbol{X}_i - \boldsymbol{X}_j\|), \qquad (2.23)$$

The straightforward calculation of the correlation integral can involve large amounts of computational time, especially for long time series, which is why

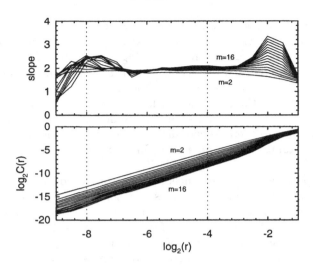

Figure 2.6: *Estimated correlation integrals (lower panel) and local slopes (upper panel) versus r for the x-variable of Rössler's model, with the embedding dimension m ranging from 2 to 16. The values between the two dotted lines were used in a least squares fit of the correlation dimension D_2 and the correlation entropy K_2 (see Figure 2.7).*

in practice one often uses efficient algorithms (for a box-assisted algorithm, see e.g. Grassberger, 1990).

For time series with relatively long decoherence times, the estimated correlation integrals may be biased as a result of the dependence among reconstruction vectors that have their time indices close. This bias can be easily corrected for in practice by excluding contributions from pairs of points that are close in time. In this way a more general estimator

$$\widehat{C}_m(r) = \frac{1}{\binom{N-T+1}{2}} \sum_{i=1}^{N-T} \sum_{j=i+T}^{N} \Theta(r - \|\boldsymbol{X}_i - \boldsymbol{X}_j\|), \qquad (2.24)$$

of the correlation integral is obtained, in which $T \geq 1$ is the so-called Theiler correction (cf. Theiler, 1986).

Figure 2.6 shows the estimated correlation integrals (with Theiler correction $T = 100$) for a time series of length $L = 10^4$ consisting of the x-variable of the Rössler model with sampling time $\Delta t = 0.5$t.u. Note that the scaling described by Equation (2.22), which implies that the correlation integrals consist of straight lines with slope D_2, appears to hold only for intermediate values

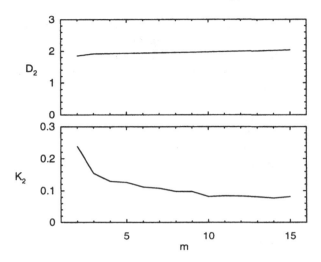

Figure 2.7: *Estimated values of the correlation dimension D_2 and the correlation entropy K_2 as a function of the embedding dimension m, based on the correlation integrals shown in Figure 2.6. The values of D_2 and K_2 appear to saturate at the values 2.0 and 0.08 nats/t.u. respectively.*

of r. The fact that it breaks down for larger values of r can be understood qualitatively by noting that the reconstructed attractor is bounded, so that there is a maximum inter-point distance above which the correlation integral is equal to 1. For small r, statistical fluctuations dominate and the estimated correlation integrals are not very reliable. For experimental time series the presence of noise also distorts the correlation integrals for small r. This will be discussed in more detail in Chapter 3.

The next step in the Grassberger-Procaccia method is the estimation of the correlation dimension and the correlation entropy from the sample correlation integral. Although strictly speaking the scaling law holds only in the limit $r \to 0$, in practice one tries to identify a so-called *scaling region* consisting of an r-interval in which the scaling law holds approximately; the curves $\widehat{C}_m(r)$ versus r on a double logarithmic scale should be approximately straight and parallel for consecutive values of m. The invariants D_2 and K_2 are estimated by fitting the scaling relation given in Equation (2.22) to the sample correlation integral within the scaling region. Of course, this procedure is not entirely objective, and estimation of D_2 and K_2 is often as much an art as it is a science.

For reliable estimation of D_2 and K_2 the embedding dimension m should be sufficiently large. Here a problem appears to arise because the minimum value of the embedding dimension is often not known. In fact, it can be shown (Sauer and Yorke, 1993) that $m > D_2$ is sufficient for the estimation of D_2. It is common practice to estimate D_2 for increasing values of the embedding dimension m and examine whether the estimated values of D_2 and K_2 saturate for increasing m. If this is the case, it may be regarded as a verification *a posteriori* that sufficiently large embedding dimensions were used. Figure 2.7 shows the estimated values of D_2 and K_2 for a time series consisting of the x-values of the Rössler model as a function of the embedding dimension. The correlation dimension D_2 saturates at a value of about 2.0 with increasing embedding dimension and the correlation entropy K_2 saturates at approximately 0.08 nats/t.u.

Sample Properties of D_2 and K_2   The sample properties of estimators of D_2 and K_2 were examined by several authors. An excellent review of the theory and estimation of fractal dimensions can be found in the paper by Cutler (1993). As noted by Cutler, the estimation of D_2 and K_2 can be considered a double limit problem; when the length of the time series increases, smaller values of r, where the scaling relations hold more accurately, should be considered. Takens (1985) proposed a maximum likelihood method for the estimation of D_2 based on the distribution of inter-point distances in the sample reconstruction measure. The dependence of estimators of D_2 and K_2 on the delay τ was studied by Caputo and Atten (1986, 1987).

Obtaining standard errors for dynamical invariants is difficult, notably as a result of the dependence among reconstruction vectors. Often one assumes that all distances smaller than some upper distance ϵ can be considered as being independent. Standard errors of estimators of D_2 are discussed by Takens (1985), Theiler (1990), Isliker (1992) and Theiler and Lookman (1993), while Olofsen *et al.* (1992) and Schouten *et al.* (1994b) discuss standard errors of maximum likelihood estimators of K_2. Approaches which take into account the dependence among delay vectors were described by Denker and Keller (1983). and Cutler (1993).

Choice of embedding delay   Although the correlation dimension and the correlation entropy of a reconstructed dynamical system are independent of the choice of the delay τ used in the reconstruction, the estimates for finite time series depend on this choice. Clearly, if a very small delay is used, the consecutive entries in a reconstruction vector hardly add any new information

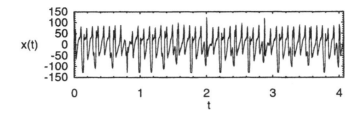

Figure 2.8: *Just over 4 seconds of dog electrogram during atrial fibrillation.*

about the state of the system. On the other hand, if the delay is taken too large, consecutive entries in a reconstruction vector will be effectively independent.

Several, mostly heuristic, choices for the delay have been proposed. One is the first zero-crossing of the autocorrelation function (provided that it has at least one zero-crossing). The idea is to minimize the linear correlation between two consecutive coordinates. Alternatively, one may choose the delay which corresponds to the first local minimum in the *mutual information* function (Fraser and Swinney, 1986) of the time series and a delayed copy of itself. The mutual information is a measure of nonlinear correlation, or redundancy, and choosing the delay so that mutual information is minimized, redundancy between consecutive coordinates of the reconstruction vectors will be small. This choice does not necessarily minimize the total redundancy in m-dimensional reconstruction vectors. Approaches based on achieving this more sophisticated goal were proposed and generally appear to imply that smaller delays should be used for larger values of the embedding dimension (see Buzug and Pfister, 1992). In fact, it turns out that it is not the delay τ, but the time window $(m - 1)\tau$ which is the most important parameter in a reconstruction. As a rule of thumb, the time window of the reconstruction should be approximately equal to one or two characteristic periods of the time series.

3.2. Example: Atrial Fibrillation

Atrial fibrillation is a commonly encountered arrhythmia of the heart. An important issue in clinical cardiology is the understanding of its physiological mechanism. Here we consider an example of an experimental time series recorded from the atrium of a chronically instrumented conscious dog in which fibrillation is electrically induced (for experimental details see Rensma *et al.*, 1988). Figure 2.8 shows the time series which has a length of just over 4 seconds and a sampling time Δt of 0.001 seconds. A phase portrait of the time series is shown in Figure 2.9 for a delay time of 21 sampling intervals

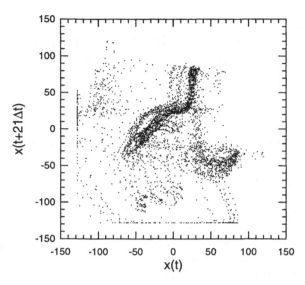

Figure 2.9: *Phase portrait (delay time 0.21s.) for the atrial fibrillation time series shown in Figure 2.8.*

(which corresponds to the observed first local minimum in the mutual information function). Figure 2.10 shows the estimated correlation integrals and their local slopes for embedding dimensions ranging from 2 to 16. The delay used in the reconstruction is 21 sampling intervals while the Theiler correction T was set to 42 sampling intervals. A clear scaling behavior is apparent for $r \in [0.0625, 0.25]$.

Figure 2.11 shows the estimated values of D_2 and K_2 for increasing values of the embedding dimension. The estimate $\widehat{D}_2$ appears to saturate at a value of about 2.6 above embedding dimension 10, and $\widehat{K}_2$ at 10 nats/s. The small increase of the dimension estimate after embedding dimension 10 may be attributed to the presence of a small amount of measurement noise. In fact, in Chapter 6 we will estimate D_2 and K_2 again for this time series, using a model that accounts for Gaussian measurement noise. An increase in the estimate of D_2 with the embedding dimension will then no longer be observed.

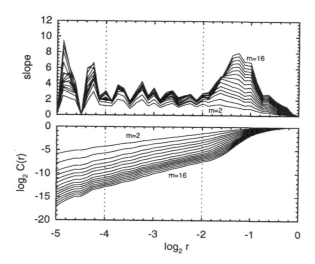

Figure 2.10: *Correlation integrals and local slopes for embedding dimensions $m = 2, \ldots, 16$ for the atrial fibrillation time series.*

4. Other Invariants

4.1. Generalized Dimensions and Entropies

As a generalization of the correlation integral, we define the order-q correlation integral $C_{q,m}(r)$ at embedding dimension m as

$$(C_{q,m}(r))^{q-1} = \int \left(\int \Theta(r - \|x - y\|) \, d\mu_m(y) \right)^{q-1} d\mu_m(x), \qquad (2.25)$$

with $q \in \mathbf{R}$. For $q = 2$ one obtains the usual correlation integral. For integer values of q larger than or equal to 2, the right hand side of Equation (2.25) has an interpretation similar to the usual correlation integral; it is the probability that all points in a μ_m-random $(q-1)$-tuple have a distance smaller than r to a μ_m-random point. For these values of q the generalized correlation integral can be estimated in a similar way as $C_{2,m}$, involving q-tuples of points rather than pairs of points.

The scaling relation

$$C_{q,m}(r) \sim r^{D_q} e^{-m\tau K_q}, \qquad (2.26)$$

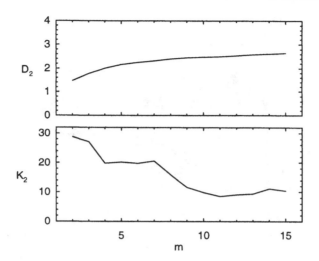

Figure 2.11: *Estimated values of the correlation dimension D_2 and the correlation entropy K_2 as a function of the embedding dimension m for the atrial fibrillation time series.*

defines the spectrum (Renyi, 1971) of Renyi dimensions D_q and entropies K_q. It was shown by Ott *et al.* (1984) that the Renyi spectrum of dimensions, D_q, is preserved under smooth coordinate transformations.

Upon taking the limit $q \to 1$ Equation (2.25) gives

$$\ln C_{1,m}(r) = \int \ln \left(\int \Theta(r - \|\boldsymbol{x} - \boldsymbol{y}\|) \, \mathrm{d}\mu_m(\boldsymbol{y}) \right) \, \mathrm{d}\mu_m(\boldsymbol{x}) \qquad (2.27)$$

The invariants D_1 and K_1 are referred to as the *information dimension* and *information entropy* respectively, and D_0, which is the fractal dimension of the support of the attractor, is known as the *box-counting dimension* or *capacity dimension* (Lasota and Mackay, 1994; Ott *et al.*, 1994).

A dynamical system is chaotic if $K_1 > 0$. Both D_q and K_q are monotonous non-increasing functions of q (see e.g. Hentschel and Procaccia, 1983). As a result, a positive value of K_2 implies a positive value of K_1, so that positive values of K_2 imply chaos.

Measures for which D_q varies with q are called *multi-fractals*, whereas if D_q is constant as a function of q, one speaks of a *monofractal*. We refer to the papers by Halsey *et al.* (1986), Paladin and Vulpiani (1987) and Peinke *et al.* (1992) for the theory of multifractals, and to those by Badii (1989), Ghez and

Vaienti (1992b) and Veneziano *et al.* (1995) for the estimation of the Renyi spectra of dimensions and entropies. Grassberger (1988) pointed out that the estimation of the generalized order-q correlation integrals involves finite sample size corrections (see also Herzel *et al.*, 1994).

Apart from the Renyi spectrum of dimensions, other definitions of the dimension of measures exist. Among those, the Hausdorff dimension D_H is best known. Usually, the information dimension D_1 and the Hausdorff dimension coincide. Counterexamples can be constructed (Radu, 1993), although these are thought to be atypical (Ott, 1993).

4.2. Lyapunov Exponents

In chaotic evolutions nearby trajectories (on average) separate exponentially. This exponential divergence can be measured in terms of the *Lyapunov exponent* which for a 1-dimensional map is defined as

$$
\begin{aligned}
\lambda(x_0) &= \lim_{n \to \infty} \frac{1}{n} \ln \left| \frac{d f^n(x_0)}{dx_0} \right| \\
&= \lim_{n \to \infty} \frac{1}{n} \ln \left| \prod_{i=0}^{n-1} f'(x_i) \right| \\
&= \lim_{n \to \infty} \frac{1}{n} \sum_{i=0}^{n-1} \ln |f'(x_i)|,
\end{aligned}
\tag{2.28}
$$

where the second equality follows from the chain rule for the derivatives of composite functions. The last expression in Equation (2.29) shows that the Lyapunov exponent is nothing but the mean growth rate of infinitesimal distances between trajectories along a reference trajectory through x_0. For ergodic systems the Lyapunov exponent has the same value for almost all choices of the initial conditions x_0 so that x_0 can be dropped as an argument of λ.

For maps of dimension k, say, the k Lyapunov exponents are defined as the mean logarithms of the moduli of the eigenvalues of the Jacobian of the map along a reference trajectory. It is common to use a decreasing ordering in the spectrum of Lyapunov exponents, i.e. $\lambda_1 > \lambda_2 > \ldots > \lambda_k$. For continuous-time dynamical systems, the spectrum of Lyapunov exponents is defined similarly. In the spectrum of continous-time dynamical systems there is always one Lyapunov exponent identically zero, since perturbations along the direction of the flow have a mean growth rate of zero.

If the evolution equations of a dynamical system are known, good numerical estimates of the spectrum of Lyapunov exponents are usually feasible. However, it is notoriously difficult to obtain reliable estimates of the spectrum of Lyapunov exponents for a dynamical system that is reconstructed from an experimental time series. Several methods were proposed for the estimation of the largest Lyapunov exponent from a time series (see e.g. Wolf et al., 1985) but these usually require large amounts of data and are sensitive to noise.

The next two equations involving Lyapunov exponents will be used in some of the later chapters. The Pesin (1977) identity

$$K_1 = \sum_{\substack{i \\ \lambda_i > 0}} \lambda_i \tag{2.29}$$

states that the information entropy K_1 is the sum of the positive Lyapunov exponents, so that a positive value of the largest Lyapunov exponent coincides with a positive value of K_1. For one-dimensional maps the Pesin identity implies that the Lyapunov exponent coincides with the information entropy K_1. The so-called Kaplan-Yorke dimension D_{KY} is defined in terms of the ordered spectrum of Lyapunov exponents, $\lambda_1 > \lambda_2 > \ldots > \lambda_k$, as

$$D_{KY} = j + \frac{\sum_{i=1}^{j} \lambda_i}{|\lambda_{j+1}|}, \tag{2.30}$$

where j is the largest integer such that $\sum_{i=1}^{j} \lambda_i > 0$. The Kaplan-Yorke conjecture (Kaplan and Yorke, 1979) states that D_1 and D_{KY} are the same under general conditions. For some models with monofractal structure the Kaplan-Yorke conjecture was verified numerically (Russel et al., 1980).

5. Further Reading

A detailed analytic study of 1-dimensional maps can be found in the book by Collet and Eckmann (1980). For introductions to the theory of dynamical systems and for additional examples we refer to the books by Schuster (1988), Wiggins (1990), Tong (1990), Broer and Dumortier (1991), Ott (1993) and De Melo and Van Strien (1993). The review paper by Grassberger et al. (1991) is an extensive introduction to nonlinear time series analysis. The books by Cvitanović (1989), Hao (1984, 1990) and Ott et al. (1994) contain collections of both theoretically and experimentally oriented papers on nonlinear dynamics, while Parker and Chua (1989) describe various numerical algorithms for generating evolutions of dynamical systems. Some recent accounts of the applications of nonlinear time series analysis can be found in the books by Kaplan and Glass (1995) and Kantz and Schreiber (1997).

CHAPTER 3

STOCHASTIC TIME SERIES

Although the theory of dynamical systems has proved to be very useful for the analysis of deterministic nonlinear time series, empirical and experimental time series often are non-deterministic. This chapter reviews several stochastic models, some of which have a stochastic term included in the evolution equations directly, while in others (spatio-temporal models) the stochastic nature of a locally measured time series originates from the interaction of local variables with neighboring variables.

1. Noise

In the previous chapter we considered deterministic time series models, which, by definition, are noise-free. Since real experiments are never entirely noise-free, these models can only serve as a reasonable description of the generating mechanism of a time series if the noise is sufficiently small. Often the noise can not be ignored and stochastic models are called for. It is important to distinguish between two types of noise: *observational noise* and *dynamical noise*. Observational noise does not affect the evolution of the dynamical system, while dynamical noise acts directly on the state of the dynamical system, thereby influencing its evolution.

1.1. Observational Noise

Observational noise, or measurement noise, can be thought of as noise imposed upon a time series by the measurement apparatus. In the simplest case the noise is i.i.d. and additive, so that the observed time series x_n can be expressed as

$$x_n = h(y_n) + \epsilon_n, \tag{3.1}$$

where y_n is the state vector at time n, h is the measurement function, or read-out function, and ϵ_n represents i.i.d. noise.

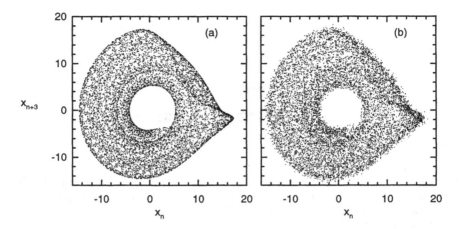

Figure 3.1: *Phase portraits for (a) a clean Rössler time series and (b) the same time series with 5% Gaussian observational noise.*

The effect of a relatively small amount of additive noise on the structure of the attractor can be clearly seen in Figure 3.1 which shows phase portraits of an x-component time series generated by Rössler's model and of the same time series with Gaussian observational noise with a standard deviation of 5% of the standard deviation of the clean time series. The fine-structure of the attractor can be observed clearly in the phase portrait of the clean time series but is no longer present in that of the noisy time series.

Measurement noise can put severe restrictions on the estimation of the dynamical invariants. It affects the correlation integral fairly dramatically, particularly for small values of r, the region of interest for estimating the dynamical invariants. Figure 3.2 shows the estimated correlation integrals of the noisy Rössler time series with 5% Gaussian observational noise. For small r the scaling behavior which was present for the clean time series (see Figure 2.6) has disappeared. The slope of the correlation integral has increased dramatically for small values of r and only a small region with approximate scaling at relatively large values of r remains visible. In fact, this phenomenon can be observed for many experimental time series. Only if the noise level is low the correlation integrals may still have a small approximate scaling region from which reasonable estimates of D_2 and K_2 can be obtained.

Noise Reduction Methods   For deterministic time series corrupted by observational noise, models for the reconstructed dynamics can be used to reduce

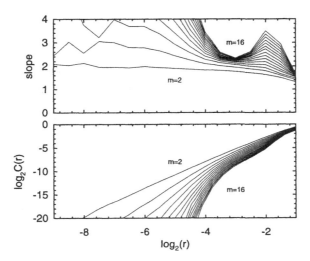

Figure 3.2: *Estimated correlation integrals (lower panel) and local slopes (upper panel) versus r for the x-variable of Rössler's model, corrupted with 5% Gaussian observational noise. The embedding dimension m ranges from 2 to 16.*

the amount of noise. Under the assumption that the observational noise is i.i.d. one may apply a nonlinear *noise reduction* scheme, which amounts to iteratively applying an adaptive nonlinear filter to the noisy time series. These approaches are based on estimating the dynamics from the noisy delay vectors reconstructed from the time series, and then adapting the observations by a small amount so that they more closely fit the estimated deterministic dynamics. Among the various methods proposed are those based on local models of the dynamics, such as locally linear models (Kostelich and Yorke, 1990) or locally constant models (Schreiber, 1993b). Sauer (1992) proposed the use of local singular value decompositions in order to distinguish locally between the model and the noise. As an alternative to local methods, Holzfuss and Kadtke (1993) proposed a global method based on radial basis functions. For reviews of noise reduction schemes we refer to Kostelich and Schreiber (1993) and Grassberger *et al.* (1993).

As argued by Mees (1993), care should be taken in using noise reduction methods as out-of-the-box tools, as they may severely disturb the structure of the time series and of the reconstructed attractor. For low noise levels, the reconstruction measures associated with the cleaned attractor should be close to the reconstruction measures of the observed time series. Generally speaking, the noise level should not be too high in order to get reliable results.

After successfully cleaning a time series by a noise reduction scheme one may proceed by applying the usual methods for the analysis of deterministic time series. However, if the primary interest lies in estimating dynamical invariants such as the correlation dimension D_2 and the correlation entropy K_2, one may skip the noise reduction step altogether and adopt a more direct approach. One can modify the Grassberger-Procaccia method in such a way that the effect of noise is included in the model of the correlation integral (see Schouten *et al.* (1994a) for bounded noise and Smith (1992b) for i.i.d. Gaussian noise). In Chapter 6 we propose a similar approach based on an adapted definition of the correlation integral which can be used in the presence of relatively high levels of Gaussian observational noise (up to 20% in standard deviation).

1.2. Dynamical Noise

A physical system, such as a stirred chemical reaction, can be described at different levels. Macroscopic evolution rules such as differential equations for chemical concentrations, provide simplified descriptions in terms of a few macroscopic variables, of systems which are strictly speaking governed by a huge number of interacting state variables, including the positions and impulses of all molecules. Note that macroscopic equations of complex systems are usually not derived directly from the microscopic equations using first principles, as this is infeasible, but found empirically. Clearly, all details on the molecular level are left out of the macroscopic equations. These can, however, have an effect on the macroscopic variables in the long run. This can be modeled by incorporating a noise term in the macroscopic equations. The properties of the noise terms and the way they enter the macroscopic equations are entirely determined by the system (Van Kampen, 1992). Typically, the level of the noise is inversely proportional to the system size and increases with temperature. This type of noise is usually referred to as *intrinsic noise*, or *system noise*. As a second type of noise influencing macroscopic variables, we mention *external noise*. One speaks of external noise if a system is subject to stochastic perturbations from the environment. Discrete time systems for which the future states depend on (possibly vector-valued) random variables ϵ_n as well as on the current state can be modeled with equations like

$$y_n = F(y_{n-1}, \epsilon_n).\tag{3.2}$$

Note that the reconstruction theorem does not apply in the presence of dynamical noise. It may even be impossible in principle to reconstruct the state of a stochastic dynamical system from an observed time series. It is conceivable that the information contained in the knowledge of the complete history track of the time series is insufficient to determine the present state of the system uniquely. Also, for time series obtained from systems with dynamical noise a clean time series is not even well defined. Formally, this implies that there is no underlying clean time series, and that noise reduction methods are inappropriate. For *hyperbolic* dynamical systems with small dynamical noise, the so-called *shadowing property* holds, which means that close to a trajectory of the noisy system there is a trajectory of the noiseless system. For these systems dynamic noise can be treated as observational noise and noise reduction may prove useful. If the shadowing property does not hold for the system under study, one can still estimate the noise-free dynamics, or 'skeleton,' by the methods proposed by Cheng and Tong (1992) and Chan and Tong (1994). It should not be expected that a time series generated by the estimated noise-free dynamics is similar to the observed time series. To reproduce the features of the original time series the skeleton should be 'clothed' with appropriate dynamical noise.

The effect of dynamical noise on the appearance of the attractor can be quite dramatic. For non-hyperbolic systems dynamical noise can lead to a so-called *noise explosion* in which the invariant measure is changed considerably. A demonstration of this phenomenon was given by Takens (1996) who found that the effect of dynamical noise on the invariant measure of the logistic map varies strongly with the nonlinearity parameter a. For certain parameter values the invariant measure is only disturbed up to a level that is comparable in size to the noise level (i.e. the disturbance is of the same order of magnitude as it would be for observational noise), whereas the invariant measure is disturbed to a much larger extend for other parameter values.

2. Stochastic Time Series Models

A brief summary of stochastic time series modeling is given here. For a review of stochastic models, including nonlinear analogs of ARMA models, we refer to the book by Tong (1990) and references therein.

Traditionally, noise sources have been considered to be the essential elements for models of erratic time series, and a large amount of theory on stochastic models is available at present. The theory on linear models is particularly well-developed and has led to a large number of linear methods for the analysis of time series, including spectral analysis (Priestley, 1981), correlation measures (Box and Jenkins, 1976) and linear system identification (Ljung, 1987).

2.1. Linear Stochastic Models

A *moving average* model MA(k) of order k is defined by

$$X_n = a_0 + \sum_{i=1}^{k} a_i \epsilon_{n-i+1}, \tag{3.3}$$

where $\{\epsilon_n\}$ represents a sequence of independent, identically distributed random variables, often called innovations. We can choose $a_0 = 0$ without loss of generality.

Autoregressive (AR) models were introduced by Yule (1927). The simplest example of autoregressive model is the first order AR(1) model

$$X_n = aX_{n-1} + \epsilon_n, \tag{3.4}$$

where $\{\epsilon_n\}$ is a sequence of i.i.d. random variables each ϵ_n being independent of all X_k for $k < n$. Often ϵ_n is referred to as the noise driving the linear autoregressive system. In order to avoid explosions of X_n to plus or minus infinity we need the condition $|a| < 1$.

Higher order autoregressive models are defined similarly. An order m autoregressive model, or AR(m) model, is given by

$$X_n = a_0 + \sum_{i=1}^{m} a_i X_{n-i} + \epsilon_n, \tag{3.5}$$

where again we may choose $a_0 = 0$ without loss of generality. The stability criterion requires that all eigenvalues of the characteristic polynomial

$$A(z) = z^k - \sum_{j=1}^{k} a_j z^{k-j} \tag{3.6}$$

are situated within the unit circle of the complex plane.

By allowing the noise term in an AR model to consist of a moving average of i.i.d. random variables, one obtains the class of *Autoregressive Moving Average* (ARMA) models. The equation for an ARMA(k, l) model is

$$X_n = a_0 + \sum_{i=1}^{k} a_i X_{n-i} + \sum_{i=1}^{l} b_i \epsilon_{n-i+1}. \tag{3.7}$$

Note that the class of ARMA models generalizes both that of MA and AR models; an ARMA$(k, 0)$ model is an autoregressive model of order k, while an ARMA$(0, l)$ model is an MA model of order l. The stability criterion and the invertibility condition require that the roots of the characteristic equations

$$
\begin{aligned}
A(z) &= z^k - \sum_{j=1}^{k} a_j z^{k-j} \\
B(z) &= \sum_{j=1}^{l} b_j z^{l-j}, \qquad b_0 = 1
\end{aligned}
\tag{3.8}
$$

of an ARMA(k, l) model are within the unit circle of the complex plane (see Tong, 1990).

Linear processes for which the invariant density is jointly Gaussian are referred to as linear Gaussian random processes (LGRPs). Their properties are determined by the autocovariance function only, and time reversal symmetry of the autocovariance function implies time reversibility of stationary linear Gaussian random processes. In Chapter 4 we will develop a test for time reversibility, which can be considered a test for a necessary property of a linear Gaussian random process, and, more generally, also of static transformation of such processes.

2.2. Nonlinear Stochastic Models

The class of linear Gaussian stochastic time series models is rich but, as noted by many authors, it is often not large enough to describe certain observed phenomena such as irreversibility. This motivates the exploration of the properties of nonlinear stochastic models.

One of the simplest classes of models beyond that of linear models is that of piece-wise linear models, or *threshold* models. In an autoregressive setup, these models have a number of linear regression coefficients in different regions of the space representing past observations. One way of selecting those regions is by the threshold principle, which gives, among others, the class of *Threshold AutoRegressive* (TAR) models proposed by Tong (1983, 1990). In the TAR(1) case, we have

$$
X_n = \begin{cases} aX_{n-1} + \sigma\epsilon_n & \text{if } X_{n-1} \leq x_{\text{th}} \\ bX_{n-1} + \sigma\epsilon_n & \text{if } X_{n-1} > x_{\text{th}}, \end{cases}
\tag{3.9}
$$

where a, b and σ are constants, $\{\epsilon_n\}$ is a sequence of i.i.d. random variables with unit standard deviation and x_{th} is the threshold value. TAR models can account for various types of behavior found in experimental time series, including irreversibility, without introducing an excessively large number of model parameters.

Polynomial models   It is often argued that polynomial nonlinear autoregressive stochastic models are non-stationary, and should therefore not be used as models for stationary nonlinear stochastic time series. Chan and Tong (1994) pointed out that this view is not fully justified and is based on the implicit assumption that the noise distribution has unbounded support. They show that under appropriate conditions systems which in the noiseless case gives rise to chaos, with dynamic noise can give rise to an ergodic stochastic system.

Sensitive dependence on initial conditions   At present there is no consensus on how the concept of a Lyapunov exponent, which characterizes the sensitivity of evolutions with respect to initial conditions, should be defined for stochastic models. Several different generalizations were proposed in the literature (see Kifer, 1986; Kapitaniak, 1990 and Yao and Tong, 1994b). For a survey of notions of initial-value sensitivity for stochastic systems we refer to Tong (1995).

Consider discrete time stochastic systems with additive dynamical noise described by

$$y_n = f(y_{n-1}) + \epsilon_n. \tag{3.10}$$

For these systems, a definition of the Lyapunov exponent analogous to that given in Equation (2.29) for deterministic systems is

$$\lambda = \lim_{n \to \infty} \frac{1}{n} \sum_{i=0}^{n-1} \ln |f'(X_i)|. \tag{3.11}$$

Herzel *et al.* (1987) noted that this Lyapunov exponent may be interpreted as the separation rate of nearby orbits under identical noise realizations. Apart from the issue of how to estimate this Lyapunov exponent from an observed time series, which is discussed by Nychka *et al.* (1992), the practical relevance of these 'same noise realization' Lyapunov exponents is not immediately obvious. If the system is in a state close to a previously visited state, its evolution will separate from the previous evolution both due to sensitive dependence on initial conditions and due to differences in the noise. As noted by Jensen (1993), identical noise realization Lyapunov exponents are not invariant under one-to-one coordinate transformations, like they are for deterministic dynamical systems. Yao and Tong (1994a) have adopted an approach based on characterizing the conditional distribution with respect to initial values, using the Kullback-Leibler information as a distance over conditional distributions given past observations. This approach leads to a sensitivity measure which has the convenient property of being invariant under smooth one-to-one coordinate transformations.

Alternatively, the discrete time stochastic dynamical systems given in Equation (3.2) can be described as a system of random maps. For these systems the states evolve according to maps which are selected randomly at each time step according to some probability distribution on a family of maps. Some authors continue along this line of thought and define the Lyapunov exponents through the derivatives of composite random maps. The Lyapunov exponent then has the form of Equation (3.11), where f is a random variable. As noted by Kifer (1986), in this approach, the non-uniqueness of the decomposition of a stochastic difference equation into a system of random maps leads to the problem of how to define Lyapunov exponents for random dynamical systems that are independent of the decomposition chosen.

3. Developments in Statistics

The theory of nonlinear dynamical systems and chaos has led to a number of developments in statistics. A survey covering all of them is beyond the scope of this book, and we will highlight only a few here. For a discussion on the connections between nonlinear dynamics and statistics we refer to Tong (1992).

Sample properties of estimators Denker and Keller (1983, 1986) proposed rigorous estimation methods for the variance of U-statistics for weakly dependent processes. An example of a U-statistic is the average of $\kappa(\boldsymbol{X}_i, \boldsymbol{X}_j)$, where $\kappa(\boldsymbol{x}, \boldsymbol{y})$ is some kernel function, over all pairs of vectors in the sample with $i \neq j$. By definition it is an unbiased estimator of $s = E(\kappa(\boldsymbol{X}_1, \boldsymbol{X}_2))$. In general, U-statistics can be constructed using kernel functions with any number of arguments (see Serfling, 1980). Note that the correlation integral for each embedding dimension m is a U-statistic based on the kernel function $\Theta(r - \|\boldsymbol{x} - \boldsymbol{y}\|)$. The statistical tests described in Chapters 4 and 5, as well as the method for estimating invariants of noisy chaotic time series (Chapter 6) all fit the framework of U-statistics.

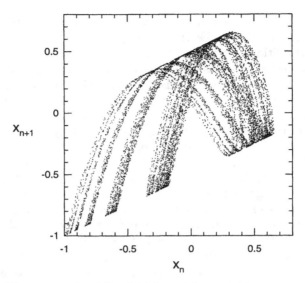

Figure 3.3: *Phase portrait of the AR(1) model $x_{n+1} = 0.5x_n + y_n$, where y_n is a time series obtained with the logistic map.*

Chaos driven AR models Linear stochastic processes are often simulated on computers using a digital random number generator in the place of a noise source. Strictly speaking, however, this noise source is deterministic and one might wonder what effect this has on the properties of estimators of linear model parameters. Motivated by this idea Stockis and Tong (1998) have examined the properties of estimators of model parameters of chaos driven linear AR models. For example, a chaos driven AR(1) model is given by

$$x_{n+1} = rx_n + \epsilon_n \qquad (3.12)$$

where $\{\epsilon_n\}$ is a chaotic process. Figure 3.3 shows a phase portrait of the AR(1) model with $r = 0.5$ with $\epsilon_n = y_n - 1/2$, where y_n evolves according to the logistic map $y_{n+1} = 4y_n(1 - y_n)$. It is well known that the classical estimators for the parameters of a linear autoregressive model are asymptotically normal if $\{\epsilon_n\}$ is an i.i.d. Gaussian sequence of random variables. It turns out that under relatively weak mixing conditions on the chaotic driving process one still obtains asymptotic normality for these estimators.

Bootstraps for dependent data For stationary mixing time series, bootstrap procedures were proposed by Carlstein (1986) and Künsch (1989). These methods account for dependence by resampling blocks of consecutive observations from the original time series, rather than individual observations. The main difference between the two methods is the way in which blocks are selected. The bootstrap of Carlstein selects from a collection of non-overlapping blocks of observations whereas the bootstrap of Künsch selects from all blocks of a given length. For a discussion on blocking rules we refer to Hall *et al.* (1995).

Nonparametric regression Nonparametric regression methods can be used to estimate the conditional distribution of a time series given past observations. Apart from being useful for the construction of nonparametric time series models, this approach may serve as the basis for alternative resampling schemes for time series with dependence. For example, a nearest neighbor resampling method was applied to river stream-flow data by Lall and Sharma (1996). The method uses information from nearby reconstruction vectors to select the next value of the bootstrap time series from the conditional distribution given past realized values of the bootstrap time series. A generalized version of this bootstrap, the so-called nearest block bootstrap, was proposed by Chan *et al.* (1997). In this method bootstrap time series are generated by concatenating blocks of observations (rather than single observations) which are again drawn from a nonparametric estimate of the conditional distribution given the past realized values of the bootstrap time series. Kreiss *et al.* (1998) consider statistical tests for simple structures such as parametric models, low order models and additivity in a nonparametric autoregressive setting, and showed that the bootstrap distribution of the test statistics asymptotically is the distribution of the statistics under the null hypothesis.

Time series with small noise can be considered operationally deterministic in that their short term behavior can be described reasonably well by a deterministic model. Yao and Tong (1998) proposed a nonparametric test for operational determinism. They examine optimal kernel widths for nonparametric kernel regression estimates of the dynamics, which should be close to zero for deterministic time series. A bootstrap method is used to establish a criterion for detecting deviations from determinism.

4. Spatio-temporal Chaos

In the previous chapter, we considered dynamical systems with a finite number of state variables together with some methods for characterizing these systems. In many spatio-temporal physical systems, however, the evolution equations consist of a set of partial differential equations governing the dependence of one or more fields in space and time, and the state space is spanned by an infinite number of state variables. It is a fundamental question as to how the methods developed for deterministic dynamical systems can be extended to these spatio-temporal situations. In some cases the theory of finite-dimensional systems readily applies to spatio-temporal systems; whenever the asymptotic evolution in the infinite-dimensional phase space is confined to a finite-dimensional attractor the theory of finite-dimensional dynamical systems is applicable. A simple example is the damped string, which asymptotically approaches its rest state. The asymptotic dynamics is described well by a stable fixed point. Similarly, in some nonlinear spatio-temporal systems the fields become organized in stable time-independent Turing type patterns (see e.g. Murray, 1989). However, the spatio-temporal dynamical structure of many other systems is of a less trivial nature, showing very complex behavior as a function of both time and space.

4.1. Coupled Map Lattices

A class of models which has recently attracted the attention of many researchers is that of *coupled map lattices*. With discrete space and time, and continuous state variables, coupled map lattices are among the simplest spatio-temporal systems. In the one-dimensional case, the evolution equations read

$$u_{n+1}^i = g(u_n^{i-1}, u_n^i, u_n^{i+1}), \tag{3.13}$$

where u_n^i denotes the state at site i and time n, and g is some function which is usually taken to be symmetric in its first and third arguments. For systems with diffusive coupling, the evolution equation has the form

$$u_{n+1}^i = (1 - \epsilon)f(u_n^i) + \frac{\epsilon}{2}(f(u_n^{i-1}) + f(u_n^{i+1})). \tag{3.14}$$

Note that this is equivalent to

$$y_{n+1}^i = f\left((1 - \epsilon)y_n^i + \frac{\epsilon}{2}(y_n^{i-1} + y_n^{i+1})\right), \tag{3.15}$$

as can be seen by substitution of $y_n^i = f(u_n^i)$, leading to

$$\begin{aligned} y_n^i &= f(u_n^i) \\ u_{n+1}^i &= (1 - \epsilon)y_n^i + \frac{\epsilon}{2}(y_n^{i-1} + y_n^{i+1}). \end{aligned} \tag{3.16}$$

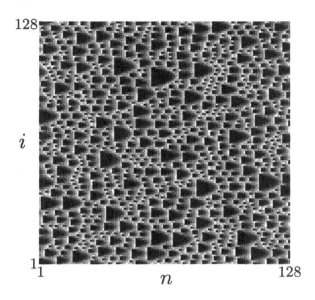

Figure 3.4: *A space-time plot of a coupled map lattice model exhibiting spatio-temporal chaos. Lighter gray scales indicate larger values of the state x_n^i at position i and time n.*

Some authors use the notation $y_n^i = u_{n+1/2}^i$ to emphasize that the evolution during one time step can be thought of as two steps. The first step consists in applying f to each of the u_n^i, and the second step in diffusion.

Figure 3.4 shows a space-time plot of a one-dimensional coupled map lattice model exhibiting spatio-temporal chaos. The model is given by Equation (3.16) with $\epsilon = 1/2$ and

$$f(u) = \alpha u \bmod 1, \qquad (3.17)$$

for $\alpha = 1.2$. The proof that this system is spatio-temporally chaotic given in Chapter 9.

Coupled map lattices are often used as metaphors for complex nonequilibrium phenomena such as convection (Yanagita and Kaneko, 1993), Rayleigh-Bénard turbulence (Jackson and Kodogeorgiou, 1992) and various other nonlinear systems, such as ecosystems (Solé *et al.*, 1992). They can display a broad variety of behaviors, including pattern formation, space-time periodic behavior, kink formation, spatio-temporal intermittency and fully developed spatio-temporal chaos. For kink formation we refer to Kaneko (1985a), and for

spatio-temporal intermittency to Kaneko (1985b) and the book by Manneville
(1990). An overview of coupled map lattices is given by Kaneko (1992).

4.2. Excitable Media

Another class of models which has recently attracted the attention of many
researchers is that of *excitable media*. These are currently being used as models
for chemical reactions, electrical activity in heart tissue, spreading of diseases,
bacteria aggregation, and galaxy formation, to mention only a few. An excitable media is a spatial system with a global stable equilibrium state, or
rest state. It can be driven out of this state, or excited, by external forces or
by interaction with neighboring excited parts of the media. The latter mechanism enables waves of excitation to propagate into un-excited regions. In
many models of excitable media, the excitation inhibits itself after some time,
so that previously excited parts of the media return to the rest state.

The simplest models of excitable media consist of two fields governed by a
set of partial differential equations,

$$
\begin{aligned}
\frac{\partial u}{\partial t} &= F(u,v) + D\Delta u \\
\frac{\partial v}{\partial t} &= G(u,v),
\end{aligned}
\tag{3.18}
$$

where $u(x,y)$ is a fast activator variable, $v(x,y)$ a slow inhibitor variable, D a
diffusion constant and $\Delta = \frac{\partial^2}{\partial x^2} + \frac{\partial^2}{\partial y^2}$ the Laplace operator.

The local dynamics (the dynamics at a fixed position (x,y) in the absence
of coupling, i.e. the ordinary differential equation obtained with $D = 0$) has
a stable equilibrium state, called the rest state, which can be assumed to be
the state $(u,v) = (0,0)$ without loss of generality. The rest state is stable with
respect to small perturbations, but when u is perturbed to values that exceed
the threshold value, an excitation is triggered in which u quickly rises to its
maximum. Typically, after an excitation of the fast activator variable u, the
slow inhibitor variable v starts growing until it has reached a certain level and
induces a quick decrease of u after which the system approaches the rest state
$(u,v) = (0,0)$ again. A typical trajectory of the local dynamics is shown in
Figure 3.5 for the model of Barkley (1991), which is obtained by taking F and
G to be

$$
\begin{aligned}
F(u,v) &= \frac{u}{\epsilon}(1-u)(u - \frac{v+b}{a}) \\
G(u,v) &= u - v.
\end{aligned}
\tag{3.19}
$$

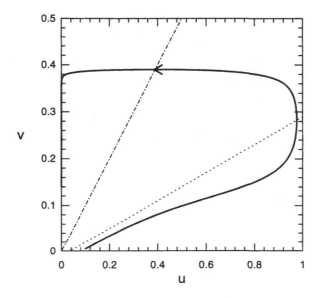

Figure 3.5: *Example of a response to a perturbation from the stable state* $(0,0)$ *in the local dynamics of Barkley's model, where u is the fast activator variable and v the slow inhibitor variable. The parameter values used are* $a = 0.3$, $b = 0.01$, $\epsilon = 1/70$. *The initial state in the u-v plane is* $(0.1, 0.005)$. *The dotted line is a null-cline of the fast activator variable u. In the parameter region below this line u increases, and above it u decreases. It intersects the u-axis at a small positive value of u. Other null-clines of u are the lines* $u = 0$ *and* $u = 1$. *The line* $u = v$, *drawn dashed-dotted, is the null-cline of the slow inhibitor variable v. In the region on the right of this line v increases, and on the left of it v decreases. After an excitation the state returns to the stable rest state* $(0,0)$ *along one of the null-clines of u (the line* $u = 0$*).*

The lines in the u-v plane on which $F(u, v) = 0$ and those on which $G(u, v) = 0$ are called the *null-clines null-cline* of u and v respectively. The threshold value for u is roughly b/a (for small v) whereas the small parameter ϵ controls the speed ratio of v and u. As a result of the diffusion of the fast activation variable, regions in the x-y plane that are in the excited state may trigger waves of excitations by forcing u above its threshold value in their neighborhoods.

Roughly speaking, four types of behavior of the activation patterns are exhibited by excitable media: plane waves, focal waves, spiral waves, and complex spatio-temporal behavior. Plane waves are waves that simply run from one side of the medium to the other. In heart tissue, these are the waves associated with the normally functioning heart. In focal waves an annulus-

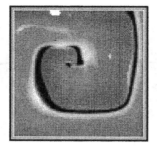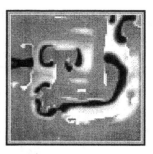

Figure 3.6: *Spiral wave break-up in the model of Panfilov and Hogeweg. The initial spiral wave is shown in the left panel. The middle panel shows the spiral wave just after breakup. After the breakup, the model evolves into complex behavior (right panel). In these figures, black corresponds to the excited state, dark gray to the absolute refractory state and the lighter grey levels to different levels of relative refractory state.*

shaped wave of excitation spreads outwards from a point of local stimulation of the medium that initially was in the rest state. Rotating spiral wave solutions can be induced in nearly all models of excitable media. The tip of the spiral wave may have a fixed position, or may wander across the medium. If the spiral tip moves around in a circular manner, the spiral is said to exhibit *simple rotation*. More complex behavior of the tip is *meander*, which can exist, for example, of two-frequency *compound rotation* of the tip. If the path of the center is more complicated one speaks of *hyper-meander*.

The study of excitable media is of particular importance for obtaining insight in the means by which *atrial fibrillation* is initiated and sustained on the atria. Since the pioneering work of Moe *et al.* (1964), who described atrial fibrillation as multiple random wavelets running across the atria, a large number of models for atrial fibrillation have been proposed. By now, atrial fibrillation is thought to correspond to complex spatio-temporal behavior which is possibly spatio-temporally chaotic. Atrial fibrillation is erratic both in space and in time and involves a number of wavelets which interact with one another in a complex manner. Models of the excitable media type are the most recent models of atrial fibrillation and appear to capture many of the phenomena observed in real atria. Although it was initially believed that excitable media should be inhomogeneous (e.g. contain regions with reduced excitability) in order to produce fibrillation-like space-time behavior, Panfilov and Hogeweg (1993) introduced a model in which this behavior results from the spontaneous break-up of a spiral wave in a homogeneous model of the atrium. Three snapshots showing the activation patterns in this model at different times are shown in Figure 3.6, illustrating the breakup of a spiral wave. Other models of the atria showing spiral wave break-up into complex spatio-temporal behavior have been described by Bär and Eiswirth (1993) and Karma (1993).

The most important issue concerning atrial fibrillation at this early stage in modeling atrial fibrillation appears not to be which of these models is the right one; at present, all these models have to be regarded as metaphors reproducing only some aspects of the phenomenology of atrial fibrillation. Eventually, the choice of a model should be based upon features that are considered to be the most relevant, but it is not yet clear which these are. For example, some mathematical idealizations that appear to have only minor effects on the local dynamics, may well turn out to be relevant for the specific way in which the spatio-temporal complexity is initialized and sustained in real atria. This is a strong argument for the development of characterization methods for spatio-temporal dynamical systems, not only within the realm of atrial fibrillation but for spatio-temporal dynamical systems in general.

4.3. Characterization of Spatio-temporal Chaos

At present little is known concerning the characterization of spatio-temporal complexity. One can prove that the evolution described by an autonomous set of partial differential equations in a finite region of space asymptotically is confined to a finite-dimensional attractor (Debussche and Marion, 1992). Formally, the reconstruction theorem applies in these cases, and it is possible to reconstruct the attractor of the system from a single time series (e.g. a locally measured time series, or a global time series consisting of an average of an observable over space). The dimension of the attractor, however, may be far too high for estimation with the chaos analysis methods currently available, which can be applied for estimation of dimensions up to about 5.

In fact, the dimension of the attractors of spatio-temporally chaotic systems is conjectured to be an extensive quantity, growing linearly with the system size. The argument is that, for large systems, regions far apart are only weakly coupled and behave practically independently. This has prompted some authors to introduce intensive quantities like the dimension density, $d_1 = D_1/V$, where V is the volume of the system. The inverse of the dimension density is a volume $v_1 = 1/d_1$ which can be thought of as the volume of a unit dimension. If the system size V is small (of the order of a few times v_1), then the system can be expected to exhibit low-dimensional behavior and chaos analysis could be used, whereas if V is much larger than v_1, traditional chaos analysis is practically useless. The review paper by Cross and Hohenberg (1993) provides the following picture. The complexity of a locally measured time series depends on the size of the system, and if the system is small compared to some typical volume, the dynamics in the infinite-dimensional phase space is confined to a low-dimensional attractor. However, if the system is large, distant regions influence the evolution of the local state variable by the propagation of disturbances through the media. The system is considered as a large number of chaotic cells, being weakly coupled and thus long-range independent. Because the dynamics in each cell shows sensitive dependence on initial conditions small local disturbances can have large global effects in the long run.

The concepts of dimension density and entropy density are promising quantities for the characterization of spatio-temporal dynamics, and it is of importance whether these quantities can be estimated from a locally measured time series. The methods proposed in the literature for the determination of dimension and entropy densities (Bauer *et al.*, 1993; Tsimring, 1993) are based on the analysis of a single time series. However, the dimension and entropy densities of a system can not always be obtained from a single time series. An example where this is not possible is the solvable model presented in Chapter 9 which produces similar time series and spatial series for parameter values for which the model has different values of the entropy density. It is not known whether these examples are of a generic nature, but at least they suggest that determining dimension and entropy densities from a single time or spatial series is not always possible. Although several multivariate nonlinear time series analysis methods are available at present, they are usually based on the assumption of low dimensionality of the global system, and new methods are required for the analysis of high dimensional spatio-temporal chaotic dynamical systems.

In the analysis of physiological data practical problems arise because the time series typically are noisy and approximately stationary only over short periods of time. Nearly all physiological time series are irregular and for many physiological time series the nature of these irregularities is not understood very well at present. The irregularities may be both the result of noise or of nonlinear determinism (Bélair *et al.*, 1995). Similar considerations apply to the analysis of various data from physical experiments. One usually deals with a nonlinear system far from equilibrium and even if the microscopic equations are simple, the behavior of the system can be very complex. In the study of turbulence often a transition from regular to complex behavior can be observed if a control parameter is changed. The behavior just after loss of stability of the equilibrium solution can be described well by low-dimensional models. In the turbulent regimes, however, the behavior is far more complex. New scaling relations may be discovered in these regimes, such as scaling laws relating characteristics of the dynamics to the system size.

5. Stationarity and Reconstruction Measures

In the next chapters the concept of the reconstruction measure associated with a stationary time series will be used frequently. An infinitely long time series is *stationary* if its 'structure' does not change over time, or, more formally, if its probabilistic properties do not depend on the time at which one starts taking measurements. In the previous chapter we sometimes imposed conditions on the model parameters of deterministic dynamical systems. The reason for not allowing every possible value of a parameter is that we want to focus on stationary evolutions of these systems. When, for example, the parameter a in the logistic map is chosen to be larger than 4, evolutions from almost all initial states in $[0, 1]$ will wander off to minus infinity; such a behavior is evidently unacceptable for a population size. In some cases one has a fairly good idea about the type of non-stationarity in a time series, and this knowledge can be used to remove the non-stationarity. For example a linear trend on the time series can be estimated using a least squares fit procedure, upon which the trend is easily removed. In situations where non-stationarity is physically plausible, such as for a time series of the position of a Brownian particle, the time series can often be rendered stationary by taking differences. In the study of economical time series such as daily stock prices it is common to focus on the differenced logarithmic price, or daily return, rather than the actual stock price.

An issue related to stationarity is that of change point detection. There, the aim is to find points in time at which a structural change occurs in the dynamics of a system. An important goal in epilepsy research is the detection of changes in the EEG (ElectroEncephaloGram) of epileptic patients who are under constant monitoring. It is conceivable that an epileptic seizure, which usually starts in a localized part of the brain, called the focus of the seizure, gives rise to a change in dynamics of the EEG even before the seizure entrains other parts of the brain and triggers a full epileptic seizure (which involves a much larger part of the brain). The prediction of epileptic seizures from a change point in the EEG is of vital importance, provided, of course, that the development of a seizure can be detected sufficiently long in advance.

5.1. Reconstruction Measures

The correlation dimension of deterministic time series, discussed in Chapter 2, is an example of an invariant of the dynamics. At the same time, it is a scaling parameter of the underlying measure of the reconstruction vectors. Here we present a rather general view of time series analysis in terms of the characterization of delay vector measures, or reconstruction measures. Deterministic time series then turn out to be members of a special class of finite order time series with reconstruction measures that have a finite dimensional (possibly fractal) structure. Similarly, linear Gaussian random processes generate time series for which the reconstruction measures are multivariate Gaussian.

To facilitate the discussion, we start by giving a definition of stationarity that is closely related to the existence of reconstruction measures (Takens, 1996). A bounded, infinitely long time series $\{X_n\}_{n=1}^{\infty}$ is called stationary if the averages

$$\bar{g} = \lim_{n \to \infty} \frac{1}{n} \sum_{k=1}^{n} g(X_k, X_{k+1}, \ldots, X_{k+m-1}) \qquad (3.20)$$

exist for each m and each continuous function $g : \mathbf{R}^m \to \mathbf{R}$. Under this condition the Riesz representation theorem (see e.g. Feller, 1966; Lasota and Mackay, 1994) states that there is an associated probability measure μ_m on $\mathbf{R}^m$, which we will refer to as the m-dimensional *reconstruction!measure*.

Note that a two-sided definition of stationarity can be obtained similarly upon replacing the average over positive times by an average over time indices running from $-n$ to $+n$. For the applications we have in mind, however, the one-sided definition is more convenient; it is known that many chaotic systems show transient behavior during which the evolution settles down on the attractor. As we are mainly interested in the asymptotic (long term) behavior of the time series, the one-sided definition is adopted. The assumption that the time series is bounded can be relaxed by considering continuous functions g on the one-point compactification of $\mathbf{R}^m$ which has all points in $\mathbf{R}^m$ at infinity identified with a single point. In order for the time series to have a proper reconstruction measure in $\mathbf{R}^m$, the measure should assign zero probability to the point representing infinity.

Various other definitions of stationarity are used in the literature. The definitions usually depend on the context and on the aims of the authors. The most notable difference with most other definitions of stationarity is that we define stationarity for a time series rather than for the process generating the time series, as is more common in statistics (see e.g. Tong, 1990). We note that processes that are strictly stationary in the usual statistical sense generate time series that are stationary in our sense with probability one.

5.2. Probabilistic Aspects

This section briefly describes some probabilistic properties of reconstruction measures. An excellent introduction to probabilistic concepts related to stochastic processes can be found in the book by Kloeden and Platen (1995).

The m-dimensional reconstruction measures μ_m of a stationary time series are probability measures. A given subset $A \subset \mathbf{R}^m$ has a measure associated with it given by

$$\mu(A) = \lim_{n \to \infty} \frac{1}{n} \sum_{k=1}^{n} I_A(X_k, X_{k+1}, \ldots, X_{k+m-1}), \qquad (3.21)$$

where $I_A(\cdot)$ is the indicator function

$$I_A(x) = \left\{ \begin{array}{ll} 1 & \text{if } x \in A, \\ 0 & \text{if } x \notin A. \end{array} \right. \qquad (3.22)$$

One may think of $\mu_m(A)$ as the relative amount of time spend in a A by the m-dimensional reconstruction vectors on average in the long run.

Obviously, the reconstruction measures μ_m for different values of m are related by certain consistency requirements. For example, the projection of the m-dimensional reconstruction measure on the first or last $m - 1$ coordinates should give the $(m - 1)$-dimensional reconstruction measure. This implies that not all probability measures are possible reconstruction measures. A simple example of a probability measure that is not the reconstruction measure of any stationary time series is a two-dimensional probability measure which has two different marginal probability measures; for a stationary time series the marginal probability measures of both components should equal the one-dimensional reconstruction measure.

Through conditioning on the first $(m - 1)$ values of a reconstruction vector, one obtains the conditional distributions of the mth element of an m-dimensional reconstruction vector given the first $(m - 1)$ elements. The conditional measures obtained from the reconstruction measure are called $(m - 1)^{\text{th}}$ order *prediction measures*. If the order of the time series is smaller than or equal to $m - 1$, the $(m - 1)^{\text{th}}$ order prediction measures determine all higher dimensional reconstruction measures inductively.

The study of nonlinear time series can be considered the study of reconstruction measures and prediction measures of time series. This picture is consistent with chaos analysis as well as with the traditional statistical approach. For a finite dimensional deterministic time series the reconstruction measure for sufficiently large values of m, by the reconstruction theorem is confined to a low (possibly fractal) dimensional set, and the prediction measure consists of a point mass at a position which is a deterministic function of the past m observations. Estimates of the dimension of a deterministic time series are means by which the reconstruction measure is characterized and estimates of measures for the sensitive dependence on initial conditions, such as the correlation entropy or largest Lyapunov exponent of a time series involve the dependence of the prediction measure on the state of a system. Stochastic time series analysis methods can also be fitted in this picture. The estimated model parameters for the class of ARMA models, for example, completely determine the multivariate probability density function of all orders and hence the reconstruction measures of all orders. Both chaos analysis and statistical time series analysis can thus be viewed as two different means for analyzing reconstruction measures. Traditionally, the statistical approach has focused on continuous (indeed mostly Gaussian) reconstruction measures and prediction measures, while chaos analysis has been concerned with fractal reconstruction measures and point-mass prediction measures. This suggests that there is a vast area for future research in between these two extremes.

In a related spirit, Tong (1990) has given an interpretation of noisy dynamical systems in terms of Markov processes. The reconstruction measure can be considered to be the invariant measure of a Markov process, while the prediction measure describes the transition probabilities of this process. Deterministic dynamical systems can be considered as Markov processes for which the conditional transition measure is concentrated on a single state given by a deterministic function of the past observations.

6. Further Reading

The book by Kapitaniak (1990) considers continuous-time nonlinear dynamical systems with dynamical noise. For numerical solutions of stochastic differential equations, we refer to the book by Kloeden and Platen (1995). A large number of nonlinear autoregressive models are discussed in the book by Tong (1990) from a dynamical systems as well as a statistical point of view. For extensive overviews of spatio-temporal dynamical systems, we refer to Meron (1992) and Cross and Hohenberg (1993).

CHAPTER 4

A TEST FOR REVERSIBILITY

This chapter proposes a test for (time) reversibility. Linear Gaussian random processes generate reversible time series and because reversibility is preserved under nonlinear static transformations, static transformations of linear Gaussian random processes comply to the null hypothesis. The test thus can be viewed as a test for linear Gaussianity which accounts for trivial nonlinearities caused by static transformations.

1. Hypothesis Testing

The importance of tests for hypotheses regarding the nature of the generating mechanism of observed time series has become well recognized in the chaos literature since it became known that even stochastic linear time series can give rise to spurious identification of low-dimensional chaos with the Grassberger-Procaccia method. Because nonlinearity is a necessary condition for chaotic behavior, particularly the question whether an observed erratic time series is compatible with a linear random process has received attention. A test for the hypothesis that a time series consists of independent, identically distributed (i.i.d.) observations was developed by Brock *et al.* (1987). Takens (1993) proposed a parametric bootstrap test for linearity. A fitted optimal linear model is used to generate 'surrogate' time series with innovations taken from the residuals of the fit. The surrogates are then compared with the original time series using the correlation integral as a test statistic. Monte-Carlo like methods based on phase randomized surrogate time series have been advocated by several authors (Kennel and Isabelle, 1992; Price and Prichard, 1993; Theiler *et al.*, 1992a; Theiler *et al.*, 1992b). It is argued that these methods are tailored for testing the hypothesis that a time series is a realization of a linear Gaussian random process, and that an exact level-α test can be constructed based on any statistic of interest. However, Timmer (1995) has given examples of test statistics for which the phase randomization method has a size larger than the nominal level. Chan (1996) has shown that the phase randomization method for time series of finite length strictly speaking is only valid for testing the null hypothesis that a time series is a realization of a circular linear Gaussian random process, and is only asymptotically valid for non-circular linear Gaussian random processes, provided that the test statistic is asymptotically independent of edge effects. Reviews of tests for linearity developed in statistics can be found in the review paper by Tsay (1986) and in the book by Tong (1990).

Whether a time series admits a linear description depends on the measurement function used, as already a one-to-one nonlinear static transformation f, which maps each of the observations X_n of a linear Gaussian time series to $Y_n = f(X_n)$, will introduce nonlinearity in the time series $\{Y_n\}$. Some authors have accounted for this so-called 'trivial nonlinearity' by absorbing it in the null hypothesis. Theiler *et al.* (1992b) and Kennel and Isabelle (1992) for example use the phase randomization method combined with a one-to-one static transformation which transforms the marginal distribution of the time series to a Gaussian distribution. Other techniques which partly account for static transformations of time series but are not based on surrogate data have been developed by Kaplan and Glass (1992, 1993) and by Wayland *et al.* (1993). A test for reversibility based on fourth-order sample cumulants was proposed by Giannakis and Tsatsanis (1994).

We are particularly interested in developing a test that takes into account the possibility that the generating process of the time series is a nonlinear static transformation of a linear Gaussian random process. We achieve this by focusing on reversibility. A stationary time series is said to be reversible if its probabilistic properties are invariant with respect to time reversal (in which case the reconstruction measure is invariant under time reversal). Static transformations of linear Gaussian random processes are reversible even if the transformation is not one-to-one. This is one clear advantage of testing for reversibility over the distribution adapted phase randomization method, which can only handle one-to-one static transformations of linear Gaussian random processes. The test proposed here is of a conventional type, in that no surrogate data are required to determine the mean and variance of the test statistic under the null hypothesis. If, however, a Monte-Carlo like test procedure is preferred a randomization procedure can be easily used in conjunction with the test statistic.

2. Reversibility

A stationary random process $\{X_n\}$ is reversible if $(X_n, X_{n+1}, \ldots, X_{n+k})$ and $(X_{n+k}, X_{n+k-1}, \ldots, X_n)$ have the same joint probability distribution for

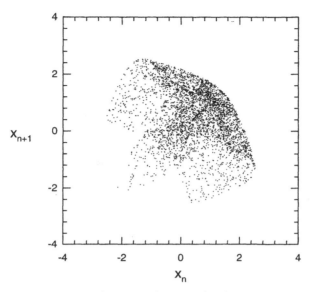

Figure 4.1: *Phase portrait (X_n, X_{n+1}) where $\{X_n\}$ is the time series obtained by adding two independent realizations of Hénon's map, one of which is reversed. The parameter values are $a = 1.4$ and $b = 0.3$. The resulting time series is reversible and has a phase portrait which is symmetric with respect to reflection in the diagonal line.*

all k and n (Lawrance, 1991). The following is an equivalent definition of reversibility in terms of reconstruction measures. A time series is reversible if the m-dimensional reconstruction measure μ associated with the reconstruction vectors

$$\boldsymbol{X}_n = (X_n, X_{n+\tau}, \dots, X_{n+(m-1)\tau}) \in \mathbf{R}^m, \qquad (4.1)$$

for all m is invariant under the transformation P, which acts on points in $\mathbf{R}^m$ as $P(x_1, \dots, x_m) = (x_m, \dots, x_1)$. Now, if we define the measure $P\mu$ to be the image of μ under P, i.e.

$$P\mu(A) = \mu(P^{-1}A) \qquad \text{for } A \subset \mathbf{R}^m, \qquad (4.2)$$

we can state that a time series is reversible if $\mu = P\mu$. It should be remarked that reversibility can also be defined for dynamical systems rather than for time series. The connection between the two definitions is not trivial and discussed to some extend in Appendix A.

The properties of linear Gaussian processes are fixed entirely by the autocorrelation function (Kaplan and Glass, 1993), and because the autocorrelation

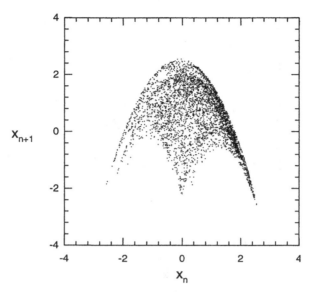

Figure 4.2: *Phase portrait* (X_n, X_{n+1}) *where now* $\{X_n\}$ *is the time series obtained by adding the same two realizations of Hénon's map as in Figure 4.1, without reversing any of the realizations. The resulting time series is irreversible and the phase portrait is not symmetric with respect to the diagonal line.*

function is symmetric with respect to time reversal, linear Gaussian random processes are reversible. We note that, for the class of linear random processes, Weiss (1975) has proved the reverse statement that a reversible linear stochastic process in general has Gaussian innovations ϵ_n. One of the exceptions is non-Gaussian i.i.d. noise which always admits the rather trivial linear description $x_n = \epsilon_n$. Of particular interest in the current context is the fact that reversibility is preserved under (not necessarily monotonous) static transformations, so that static transformations of linear Gaussian random processes comply to the null hypothesis. Notice that reversibility is also preserved under some other transformations of time series. For example, taking second differences, $Y_n = X_{n-1} - 2X_n + X_{n+1}$, is symmetrical in the time order of its indices and hence will preserve reversibility. Taking first differences, $Y_n = X_n - X_{n-1}$, is not a reversibility preserving operation.

Irreversibility is often easily spotted by eye upon examining phase portraits. For a reversible time series the points (X_n, X_{n+k}) for all k should be distributed symmetrically with respect to reflection in the diagonal line. For example, Figure 4.1 shows the phase portrait of a time series obtained by adding two independent realizations of Hénon's map with parameter values $a = 1.4$ and $b = 0.3$, after reversing one of the realizations in time. It is an example of a reversible time series which is not a static transformation of a linear Gaussian random process. The distribution clearly has a symmetric appearance with respect to reflection in the diagonal. Figure 4.2 shows a phase portrait of a time series obtained by adding the same realizations of Hénon's map without reversing one of the realizations. From the fact that the phase portrait is asymmetrical, it can be concluded that the time series is not a realization of a static transformation of a linear Gaussian random process.

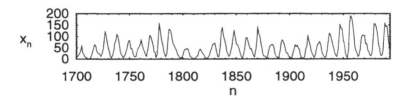

Figure 4.3: *Annual sunspot numbers (1700–1994). The up-strokes are faster than the down-strokes which implies irreversibility.*

An example of a real time series which shows signs of irreversibility, is the well-known Wolf sunspot time series shown in Figure 4.3. The irreversibility can be clearly seen in this time series; the up-strokes generally take place on shorter time scales than the down-strokes. The irreversibility of the sunspot time series is also clearly visible in the phase portrait shown in Figure 4.4. In the faster up-strokes the difference between the two delay coordinates X_{n-1} and X_n is larger which results in larger distances from the diagonal line during the up-strokes (upper-left of the diagonal) than during the down-strokes (lower-right of the diagonal).

3. Test Statistic

In this section we propose a statistic for the null hypothesis H_0 that a time series is reversible,

$$H_0 : \qquad \mu = P\mu. \tag{4.3}$$

For the moment it is assumed that the N reconstruction vectors $\{X_n\}_{n=1}^N$ are independently distributed according to the reconstruction measure μ. Of course, this assumption is not realistic for many experimental time series, and can only be expected to be reasonable for time series with short-range

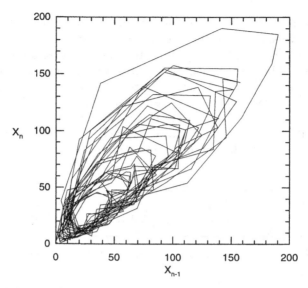

Figure 4.4: *Phase portrait of annual sunspot numbers (1700–1994). The points are connected for clarity. The asymmetry with respect to reflection in the diagonal line indicates irreversibility of the time series.*

dependence. Later in this chapter it will be discussed how the assumption of independence of the reconstruction vectors can be relaxed.

A kernel estimator tailored for the null hypothesis, H_0, can be constructed in the following way. For positive d and two finite (possibly signed) measures μ_1 and μ_2 we define the bilinear form

$$(\mu_1, \mu_2) = \int\int K(r - s)\, \mathrm{d}\mu_1(r)\, \mathrm{d}\mu_2(s) \qquad (4.4)$$

where

$$K(r - s) = e^{-\|r-s\|^2/(2d^2)}, \qquad (4.5)$$

and $\|\cdot\|$ denotes the Euclidean norm in $\mathbf{R}^m$. In fact, the bilinear form is positive definite so that (μ_1, μ_2) is an inner product of μ_1 and μ_2.

Proposition 4.1 *For two probability measures μ_1 and μ_2 on $\mathbf{R}^m$, the quantity*

$$R = (\mu_1 - \mu_2, \mu_1 - \mu_2) \qquad (4.6)$$

is zero if μ_1 and μ_2 are equal and larger than zero otherwise.

Proof: Clearly, by bilinearity R is zero if μ_1 and μ_2 are equal, and it remains to be shown that R is larger than zero if μ_1 and μ_2 differ. The convolution in Equation (4.4) allows us to rewrite R as an integral over Fourier space,

$$R = \int \hat{K}(\boldsymbol{k}) |\hat{f}_1(\boldsymbol{k}) - \hat{f}_2(\boldsymbol{k})|^2 \, d\boldsymbol{k} \tag{4.7}$$

where $\hat{f}_1$ and $\hat{f}_2$ are the characteristic functions of μ_1 and μ_2 respectively, and $\hat{K}$ is the Fourier transform of the kernel function K. Because probability measures are uniquely determined by their characteristic functions (Billingsley, 1979) and the Fourier transform of K is real and everywhere positive, R is larger than zero if μ_1 and μ_2 are not equal. $\square$

By proposition 4.1 the quantity

$$Q = (\mu - P\mu, \mu - P\mu) \tag{4.8}$$

equals zero if $\mu = P\mu$ and is positive otherwise. Now, $(\mu, \mu) = (P\mu, P\mu)$ by the fact that P is an isometry, and $(\mu, P\mu) = (P\mu, \mu)$ by bilinearity, and one obtains

$$\begin{aligned} Q &= 2(\mu, \mu) - 2(\mu, P\mu) \\ &= \iint g(\boldsymbol{r}, \boldsymbol{s}) \, d\mu(\boldsymbol{r}) \, d\mu(\boldsymbol{s}) \end{aligned} \tag{4.9}$$

with

$$g(\boldsymbol{r}, \boldsymbol{s}) = 2 \left(e^{-\|\boldsymbol{r}-\boldsymbol{s}\|^2/(2d^2)} - e^{-\|\boldsymbol{r}-P\boldsymbol{s}\|^2/(2d^2)} \right). \tag{4.10}$$

Based on Equation (4.9), a U-estimator $\hat{Q}$ of Q can be constructed by replacing the double integrals over μ by an average of contributions from different pairs of delay vectors $\boldsymbol{X}_i$ and $\boldsymbol{X}_j$:

$$\hat{Q} = \frac{2}{N(N-1)} \sum_{i=1}^{N-1} \sum_{j=i+1}^{N} W_{ij}, \tag{4.11}$$

where W_{ij} is

$$W_{ij} = g(\boldsymbol{X}_i, \boldsymbol{X}_j). \tag{4.12}$$

It can readily be checked that Equation (4.11) gives an unbiased estimator of Q by calculating the expected value of each term W_{ij} for each pair of vectors, $\boldsymbol{X}_i$ and $\boldsymbol{X}_j$, which are assumed to be drawn independently from μ, i.e.

$$E(W_{ij}) = \iint g(\boldsymbol{x}, \boldsymbol{y}) \, d\mu(\boldsymbol{x}) \, d\mu(\boldsymbol{y}) = Q, \tag{4.13}$$

so that, for any probability measure μ, we have $E(\widehat{Q}) = Q$. Furthermore, since $g(\cdot, \cdot)$ is bounded, we have $E(g^2(\boldsymbol{X}_i, \boldsymbol{X}_j)) < \infty$, which implies consistency of $\widehat{Q}$ (see Serfling, 1980), so that $\widehat{Q} \to Q$ with probability 1 for $N \to \infty$. Because $Q > 0$ for any alternative, a one-sided test based on $\widehat{Q}$ which rejects for large values of $\widehat{Q}$ has power against any fixed alternative.

The null distribution of $\widehat{Q}$ is not easily obtained, even under the assumption of independence of the reconstruction vectors. However, under this assumption, there is a minimal sufficient statistic for the class of reconstruction measures under the null hypothesis, conditionally on which the mean and variance of $\widehat{Q}$ can be calculated easily. We consider the group $G = \{e, P\}$ consisting of the identity e and the transformation P and define the G-orbit $G\boldsymbol{x}$ of a vector $\boldsymbol{x} \in \mathbf{R}^m$ as

$$G\boldsymbol{x} = \{\boldsymbol{x}, P\boldsymbol{y}\}. \tag{4.14}$$

G-orbits enable the formulation of the following proposition.

Proposition 4.2 *The sample set of G-orbits, $\{G\boldsymbol{X}_1, \ldots, G\boldsymbol{X}_N\}$, under the assumption of independence of the reconstruction vectors, and under the null hypothesis, is a minimal sufficient statistic for the class of probability measures that are invariant under P.*

Proof: The set of G-orbits is a maximal invariant under the group $\{e, P\}$, and hence a minimal sufficient statistic for the class of probability distributions that are invariant under P.   □

Because $\widehat{Q}(\boldsymbol{X}_1, \ldots, \boldsymbol{X}_N)$ is invariant under permutations of its arguments, the distribution of $\widehat{Q}$ conditionally on $\{G\boldsymbol{X}_1, \ldots, G\boldsymbol{X}_N\}$ is identical to the distribution of $\widehat{Q}$ conditionally on the ordered set of G-orbits $(G\boldsymbol{X}_1, \ldots, G\boldsymbol{X}_N)$. This shows that for the calculation of properties of the conditional distribution of $\widehat{Q}$, we may condition on the ordered set of G-orbits rather than on the unordered set of G-orbits. The conditional random process then consists of randomly selecting a point on each G-orbit, according to their conditional probabilities under the null hypothesis. Under the null hypothesis, the two different vectors associated with a G-orbit both have probability 1/2 of being selected. Note that $g(\cdot, \cdot)$ satisfies

$$g(\boldsymbol{r}, \boldsymbol{s}) = g(P\boldsymbol{r}, P\boldsymbol{s}) = -g(P\boldsymbol{r}, \boldsymbol{s}) = -g(\boldsymbol{r}, P\boldsymbol{s}), \tag{4.15}$$

which implies that conditionally on the ordered set of G-orbits the absolute values $|W_{ij}|$ are constants while each of the W_{ij} is either positive or negative, both with probability 1/2.

If we let $E_c(\cdot)$ denote expected values conditionally on the set of G-orbits under the null hypothesis then for $E_c(\widehat{Q})$ we find

$$E_c(\widehat{Q}) = 0. \tag{4.16}$$

Under the null hypothesis, the conditional variance of $\widehat{Q}$, which is $Var_c\{\widehat{Q}\} = E_c((\widehat{Q} - E_c(\widehat{Q}))^2)$ becomes

$$Var_c\{\widehat{Q}\} = \frac{4}{N^2(N-1)^2} \sum_{i=1}^{N-1} \sum_{j=i+1}^{N} \sum_{k=1}^{N-1} \sum_{l=k+1}^{N} E_c(W_{ij}W_{kl}). \tag{4.17}$$

Note that the W_{ij} are conditionally uncorrelated; if $\{i,j\} \neq \{k,l\}$ the expected value of the product $W_{ij}W_{kl}$ is zero. Suppose that one of the indices, k say, is not equal to either of the indices of the other term (in this case i and j), we have

$$
\begin{aligned}
E_c(W_{ij}W_{kl}) &= E_c(E_c(W_{ij}W_{kl}|\boldsymbol{X}_i, \boldsymbol{X}_j, \boldsymbol{X}_l)) \\
&= E_c(W_{ij}(\tfrac{1}{2}|W_{kl}| - \tfrac{1}{2}|W_{kl}|)|\boldsymbol{X}_i, \boldsymbol{X}_j, \boldsymbol{X}_l) = 0.
\end{aligned} \tag{4.18}
$$

This leaves

$$Var_c\{\widehat{Q}\} = \frac{4}{N^2(N-1)^2} \sum_{i=1}^{N-1} \sum_{j=i+1}^{N} W_{ij}^2 \tag{4.19}$$

for the variance of $\widehat{Q}$ conditionally on the set of G-orbits.

The ratio S defined by

$$S = \frac{\widehat{Q}}{\sigma_c(\widehat{Q})}, \tag{4.20}$$

with

$$\sigma_c(\widehat{Q}) = \sqrt{Var_c\{\widehat{Q}\}} = \frac{2}{N(N-1)} \left(\sum_{i=1}^{N-1} \sum_{j=i+1}^{N} W_{ij}^2 \right)^{\frac{1}{2}} \tag{4.21}$$

under the null hypothesis thus has mean zero and unit standard deviation. The theory of U-statistics shows that the asymptotic distribution of S for independent vectors is not necessarily normal (see e.g. Serfling, 1980). A derivation of the asymptotic distribution of S under the null hypothesis conditional on the set of G-orbits is beyond the scope of this book. For the example time series yet to be discussed, we will propose a method for handling dependence in the delay vector distributions. For some reversible processes the probability of finding a value of S larger than 2 and 3 respectively will be estimated numerically.

4. Simulations: Reversible Time Series

In this section the test for reversibility is applied to a number of reversible example time series and some heuristic procedures are proposed to handle the dependence among reconstruction vectors. These procedure are not rigorously justified, but they will be shown to work well for simulated time series. All time series $\{X_n\}_{n=1}^{L}$ are rescaled to have zero mean and a sample standard deviation of one. The advantage of normalizing the standard deviation rather than the range of the time series is that the standard deviation of the time series is usually less dependent on the length of the time series.

The parameter d in the kernel function given in Equation 4.5 sets the length scale of the smoothing. By choosing it relatively small, $\widehat{Q}$ will pick up differences in the local density around vectors X_i and PX_i. However, taking d too small leads to poor statistics as can be seen from the behavior of $\widehat{Q}$ in the limit $d \to 0$. If the distances $\|X_i - X_j\|$ and $\|PX_i - X_j\|$ are all different, $\widehat{Q}$ in the limit $d \to 0$ is dominated by the contribution from one pair of G-orbits only. Since small distances usually become less frequent for increasing values of the embedding dimension m, large embedding dimensions also tend to give rise to poor statistics.

For the numerical calculations we take $m = 3$ and $d = 0.2$, which turn out to be reasonable values for a broad variety of processes. For simplicity, only discrete time processes are considered here, and the delay time τ used in the reconstruction is set to 1 throughout.

4.1. i.i.d. Time Series

Dependence between delay vectors will affect the properties of the test statistic $\widehat{Q}$. In a set of delay vectors, two possible sources of dependence can be distinguished: dependence as a result of the mere construction of the delay vectors from the time series, which causes different delay vectors to have values of elements in common, and additional dependence due to the dynamics of the time series. The first source of dependence can introduce a bias in $\widehat{Q}$, even for i.i.d. time series. This will be termed a Theiler effect, as it results from the fact that contributions from pairs of delay vectors that are close in time have a different expected value than those from pairs of independent delay vectors. The resulting bias in $\widehat{Q}$, which is of the order $1/N$, can be relatively easily compensated for by incorporating a Theiler correction, which amounts to excluding contributions from pairs of points that are close in time. The test statistic with Theiler correction T is

$$\widehat{Q} = \frac{1}{\binom{N-T+1}{2}} \sum_{i=1}^{N-T} \sum_{j=i+T}^{N} W_{ij}. \tag{4.22}$$

The expected value of W_{ij} is equal to Q if $\boldsymbol{X}_i$ and $\boldsymbol{X}_j$ are mutually independent, and for i.i.d. time series and with $\tau = 1$ this is the case for all pairs $(\boldsymbol{X}_i, \boldsymbol{X}_j)$ which have $| i - j | \geq m$. It follows that Q can be estimated unbiasedly by excluding the pairs by using a Theiler correction T equal to the embedding dimension m.

The fact that the contributions of pairs (i, j) and $(i + k, j + l)$ can be correlated (either positively or negatively) for small values of k and l also affects the standard deviation of $\widehat{Q}$. This may give rise to actual type I error rates that are larger than their nominal values, that is, the test may reject too frequently under the null hypothesis. In the derivation of the standard deviation, it was assumed that all delay vectors are independent. This condition can be met for i.i.d. time series using each mth delay vector instead of all delay vectors. However, because such an approach would reduce the amount of information used in the test, it will be avoided here.

We will first examine whether the derived properties of S, viz. mean equal to zero and standard deviation equal to one, are reasonable approximations for i.i.d. time series if a Theiler correction is used, together with

$$\sigma_{\mathrm{c}}(\widehat{Q}) = \frac{1}{\binom{N-T+1}{2}} \left(\sum_{i=1}^{N-T} \sum_{j=i+T}^{N} W_{ij}^2 \right)^{\frac{1}{2}}, \tag{4.23}$$

which is the analog of Equation (4.19) for $\widehat{Q}$ with a Theiler correction. It is valid under the assumption that the W_{ij} with $|i - j| \geq T$ are uncorrelated.

We estimated the mean $\bar{S}$ and standard deviation $\sigma(S)$ of S, from S-values obtained for 100 independently generated time series. In Table 4.1 the estimated values $\bar{S}$ and $\sigma(S)$ are given for time series which are realizations of an i.i.d. process with a uniform distribution, for various time series lengths L and values of the Theiler correction T. The standard error of the estimated mean $\bar{S}$ is about $\sqrt{(1/100)}\sigma(S) = 0.1\sigma(S)$. Under the null hypothesis and for independent delay vectors $\sigma(S)$ is equal to 1. The standard error of the estimated value of the standard deviation $\sigma(S)$ is about 0.07, under the assumption that S has a Gaussian distribution.

T	$L = 50$		$L = 100$		$L = 200$		$L = 400$	
	$\bar{S}$	$\sigma(S)$	$\bar{S}$	$\sigma(S)$	$\bar{S}$	$\sigma(S)$	$\bar{S}$	$\sigma(S)$
1	−0.78	1.12	−0.87	0.95	−0.81	1.06	−0.87	1.12
2	−0.57	1.01	−0.49	1.14	−0.48	1.11	−0.49	1.24
3	0.11	1.01	0.02	1.04	0.19	1.12	−0.22	1.02
4	0.07	1.05	−0.10	0.94	0.02	1.05	0.06	1.04

Table 4.1: *Mean and standard deviation of S for i.i.d. uniformly distributed time series as estimated from 100 independently simulated time series.*

As noted above, for $T < m$, $\widehat{Q}$ will have a bias due to contributions from pairs with close time indices and since this bias is of the order $1/L$ it will vanish asymptotically. However, the results shown in Table 4.1 suggest that the test statistic S has a negative bias for $T < m$ which does not decrease with the time series length. The latter results from the fact that $\sigma(S)$ is also of the order $1/L$, so that the bias in the quotient, S, will not vanish for large L. For $T \geq m$ the bias in $\widehat{Q}$ is equal to zero, and the corresponding values of $\bar{S}$ are close to zero.

T	$L = 50$		$L = 100$		$L = 200$		$L = 400$	
	$\bar{S}$	$\sigma(S)$	$\bar{S}$	$\sigma(S)$	$\bar{S}$	$\sigma(S)$	$\bar{S}$	$\sigma(S)$
1	−0.75	1.00	−0.94	0.91	−1.00	0.91	−0.97	1.04
2	−0.59	1.01	−0.66	0.98	−0.47	0.93	−0.41	1.13
3	−0.14	0.92	−0.03	0.90	0.05	1.03	0.12	1.15
4	0.02	1.07	−0.17	0.90	0.15	1.08	−0.02	0.93

Table 4.2: *Mean and standard deviation of S for i.i.d. Gaussian distributed time series as estimated from 100 independent time series.*

The results for i.i.d. Gaussian time series, shown in Table 4.2, are similar to those found for the uniform i.i.d. time series; for $T < m$ there is a bias, and for $T \geq m$ there is no bias. For both uniform and Gaussian time series, the estimated standard deviations of S are close to 1, which suggests that the covariance of the terms W_{ij} and W_{kl} (with indices more than m apart) that contribute to $\widehat{Q}$ is negligible for the i.i.d. time series considered here.

4.2. Reversible Time Series with Dependence

Next we consider some processes that generate reversible time series with dependence which decays to zero in the course of time, or weakly dependent

T	$a = -0.9$		$a = -0.5$		$a = 0.5$		$a = 0.9$	
	$\bar{S}$	$\sigma(S)$	$\bar{S}$	$\sigma(S)$	$\bar{S}$	$\sigma(S)$	$\bar{S}$	$\sigma(S)$
1	-1.29	0.89	-0.92	1.02	-1.80	0.84	-1.76	0.74
2	-1.04	0.86	-0.69	1.12	-1.44	0.75	-1.31	0.91
3	-0.15	0.86	-0.06	0.99	-0.51	0.86	-0.50	0.96
4	-0.27	0.86	0.10	1.02	-0.20	0.78	-0.23	0.86
5	0.11	0.70	0.04	1.01	-0.27	0.77	-0.24	0.72
10	0.06	0.90	-0.02	1.10	-0.09	0.86	-0.23	0.77
15	0.00	0.87	0.01	1.12	-0.07	0.95	-0.08	0.85

Table 4.3: *Mean and standard deviation of S for AR(1) processes $X_n = aX_{n-1} + \epsilon_n$ with $\epsilon_n \sim N(0,1)$, as estimated from 100 independently simulated time series of length $L = 100$.*

time series. As in the case of i.i.d. time series, $\widehat{Q}$ may be biased due to the contributions from G-orbits W_{ij} which have close time indices i and j, but for sufficiently large values of the Theiler correction T this bias should be negligibly small.

Table 4.3 shows the estimated values of $\bar{S}$ and $\sigma(S)$ for an AR(1) process $X_n = aX_{n-1} + \epsilon_n$, with $\{\epsilon_n\} \sim N(0,1)$, for the parameter values $a = -0.9, -0.5, 0.5, 0.9$. The time series have length $L = 100$ and again the test parameters $m = 3$, $\tau = 1$ and $d = 0.2$ are used. The estimated bias of S decreases (in absolute value) with increasing values of the Theiler correction. The bias of S is the largest for $a = 0.9$, a case with relatively slowly decaying dependence. The standard deviation of S is reasonably close to 1 for $a = \pm 0.5$ but appears to be too small for $a = \pm 0.9$. The small standard deviations of S suggest that the standard deviations of $\widehat{Q}$ are overestimated, as a result of which the test will be conservative (rejects less often under the null hypothesis). An improved estimator of the standard deviation of S will be discussed later. We conclude that, at least for the AR(1) processes examined, the Theiler correction substantially reduces the bias of the test statistic S, and that the variance of the test statistic is fairly robust with respect to dependence.

The next process considered is the non-monotonic static transformation of a first order linear Gaussian random process,

$$X_n = \tanh^2(Y_n), \qquad \text{where } Y_n = aY_{n-1} + \epsilon_n. \tag{4.24}$$

The estimated values of the mean and standard deviation of S for this process are given in Table 4.4 for the same values of a as examined for the AR(1) process. Again the bias decreases in absolute value with T, is now of comparable magnitude for all parameter values examined. The standard deviation is close

T	$a = -0.9$ $\bar{S}$	$\sigma(S)$	$a = -0.5$ $\bar{S}$	$\sigma(S)$	$a = 0.5$ $\bar{S}$	$\sigma(S)$	$a = 0.9$ $\bar{S}$	$\sigma(S)$
1	-1.05	0.79	-1.01	1.09	-1.18	0.95	-1.01	0.74
2	-0.91	0.86	-0.56	0.99	-0.63	0.96	-1.06	0.76
3	-0.23	0.79	0.20	1.07	-0.01	1.07	-0.39	0.80
4	-0.26	0.74	0.03	1.05	-0.07	0.96	-0.24	0.70
5	-0.23	0.68	0.09	0.94	-0.21	1.00	-0.19	0.74
10	-0.08	0.82	-0.08	1.12	-0.24	1.04	-0.08	0.74
15	-0.06	0.80	0.19	1.02	-0.06	1.00	0.15	0.74

Table 4.4: *Mean and standard deviation of S for a static transformation of the AR(1) process $X_n = aX_{n-1} + \epsilon_n$ with $\epsilon_n \sim N(0,1)$, as estimated from 100 independently simulated time series (L = 100).*

to one for the parameter values $a = -0.5$ and $a = -0.5$, while the standard deviation is smaller than one for $a = -0.9$ and $a = 0.9$.

Figure 4.5 shows the estimated mean values (bias) of $W_{(i)(i+k)}$ for some of the reversible processes examined so far, as estimated from single realizations of length 400. For small values of k a negative bias can be observed. For i.i.d. time series the expected value of $W_{(i)(i+k)}$ equals Q exactly for $k \geq m$, but for time series with dependence the bias can be expected to disappear only for larger values of k. In practice, similar figures can be used to guide the choice of the Theiler correction T, which should be chosen at least as large as the smallest value of k for which the bias of $W_{(i)(i+k)}$ is negligible.

The fact that the bias is negative for small values of k can be understood for the i.i.d. time series for small values of the bandwidth. We have

$$W_{(i)(i+k)} = e^{-\|\boldsymbol{X}_i - \boldsymbol{X}_{i+k}\|^2/(2d^2)} - e^{-\|\boldsymbol{X}_i - P\boldsymbol{X}_{i+k}\|^2/(2d^2)}. \qquad (4.25)$$

For $m = 3$ and $k = 1$, for example, the distance in the first term reads $\|(X_i - X_{i+1}, X_{i+1} - X_{i+2}, X_{i+2} - X_{i+3})^T\|$ which is smaller than some given value ϵ with a probability proportional to ϵ^3, so that the first term scales as d^3. The vector in the second term is $\|(X_i - X_{i+3}, X_{i+1} - X_{i+2}, X_{i+2} - X_{i+1})^T\|$ and because the last two elements are the same apart from their sign, this second term is smaller than ϵ with a probability proportional to ϵ^2, so that the second term for small d behaves as d^2. Therefore, the second term will always dominate for small values of d, so that $W_{(i)(i+k)}$ will be negative. Similar arguments show that for i.i.d. time series and for general values of m, the bias of $W_{(i)(i+k)}$ will be negative for small d when $1 \leq k < m$.

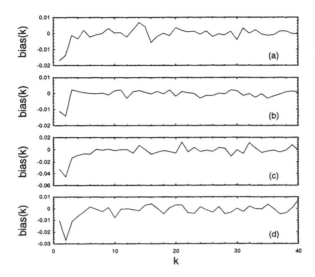

Figure 4.5: *Estimated mean of $W_{(i)(i+k)}$ as a function of k for (a) i.i.d. uniformly distributed time series, (b) i.i.d. Gaussian time series, (c) an AR(1) time series with regression parameter $a = 0.9$ and (d) a static transform (tanh) of a realization of the same AR(1) process. The number of observations in each time series is $L = 400$.*

4.3. The Block Method

Next we examine time series obtained by adding time-forward and time-backward Hénon time series, which are examples of reversible time series with dependence but not realizations of static transformation of a linear Gaussian random process. Each of the realizations is obtained by adding the x-components of two independently generated Hénon time series, after reversing one of the series. A delay phase portrait for this process was shown in Figure 4.1. Table 4.5 (first column) gives the estimated mean and standard deviation of S for this process (again for a time series length L of 100 and test parameters $m = 3$, $\tau = 1$ and $d = 0.2$). Note that the standard deviation of S is larger than 1, which suggests that the test may have an increased type I error rate for this process (rejects too often under the null hypothesis).

The large standard deviation is the result of covariance among delay vectors. This covariance can be accounted for by dividing the i-j grid of indices in

T	$\ell = 1$ $\bar{S}$	$\sigma(S)$	$\ell = 2$ $\bar{S}$	$\sigma(S)$	$\ell = 4$ $\bar{S}$	$\sigma(S)$	$\ell = 8$ $\bar{S}$	$\sigma(S)$
1	-1.76	1.14	-1.52	1.03	-0.91	1.04	-0.57	1.05
2	-1.28	1.18	-0.99	1.04	-0.79	0.99	-0.37	1.01
3	-0.75	1.03	-0.69	0.96	-0.47	1.00	-0.31	0.91
4	-0.32	1.17	-0.27	1.08	-0.21	0.83	-0.08	1.07
5	-0.27	1.17	-0.21	0.96	-0.10	0.95	-0.01	0.97
10	-0.26	1.08	-0.10	1.12	-0.07	1.09	0.04	1.14
15	0.12	1.18	-0.09	1.05	-0.15	0.96	-0.14	1.04

Table 4.5: Mean and standard deviation of S for the sum of two independent realizations of the Hénon map, one of which is reversed, as estimated from 100 independently simulated time series $(L = 100)$.

squares of size $\ell \times \ell$ and using the average value

$$W'_{i'j'} = \frac{1}{\ell^2} \sum_{p=1}^{\ell} \sum_{q=1}^{\ell} W_{(i'\ell+p)(j'\ell+q)}, \qquad (4.26)$$

in each of these squares rather than the individual values W_{ij}. The expression for the conditional variance of $\widehat{Q}$ in terms of W' is analogous to Equation (4.21). The idea behind this procedure is that for ℓ large enough, non-overlapping segments, or blocks, of length ℓ within the series of delay vectors for all practical purposes can be considered as independent. Heuristically, in this way one can approximately take into account short-range (within a time scale ℓ) dependence among delay vectors. In the following, the method with $\ell > 1$ will be referred to as the block method. The parameter ℓ in Table 4.5 refers to the block length; for $\ell = 1$, the original test statistic with Theiler correction T is obtained. The results for sums of independent Hénon time series and reversed Hénon time series are given in Table 4.5 for several values of ℓ. Indeed, with increasing values of ℓ, the standard deviation of S decreases and becomes close to 1.

Summarizing, for dependent time series it is advisable to use a Theiler correction to reduce the bias as well as a sufficiently large block size to account for the effect of covariance among contributions to the statistic from different pairs of reconstruction vectors. Good results are usually obtained when both T and ℓ are of the order of the typical time scale on which dependence decays in the time series.

4.4. Critical Levels

Given the above procedure to obtain S, one would like to have an estimate of the probability $P(S > s_r)$ of finding a value of S larger than some given value s_r. Under the null hypothesis, S has zero mean and unit standard deviation. The Chebychev inequalities provide bounds for probabilities that a random variable with known mean and variance exceeds a given value. These inequalities are often inefficient in that they lead to conservative tests. For one-sided tests at a nominal level 0.05, the Chebychev inequalities give a rejection criterion $S > 5$. For specific classes of distributions often sharper inequalities can be derived. For example, for unimodal distributions, the probability of finding a value more than 3 standard deviations above the mean is smaller than 0.05 (see Pukelsheim, 1994).

a	$\bar{S}$	$\sigma(S)$	$P(S > 2)$	$P(S > 3)$
-0.9	-0.01	0.93	0.027	0.002
-0.5	0.04	1.00	0.030	0.006
0.5	-0.05	0.97	0.023	0.000
0.9	-0.04	0.84	0.016	0.001

Table 4.6: *Mean and standard deviation of S, and estimated rejection probabilities for the static transform of the AR(1) processes discussed above, as estimated from 1000 independently simulated time series ($L = 200$).*

In the numerical simulations for reversible time series only a small number of values of S were found to be larger than 2 or 3. In order to estimate the probabilities $P(S > 2)$ and $P(S > 3)$ of finding S larger than 2 and 3 respectively, we performed an additional 1000 simulations with $L = 200$ for the static transformations of AR(1) processes discussed above. The estimated probabilities $P(S > 2)$ and $P(S > 3)$ with $T = \ell = 10$ are given in Table 4.6, as well as the estimated mean and standard deviation of S. For comparison, for the standard normal distribution, the probabilities of finding values larger than 2 and 3 are approximately 0.025 and 0.005 respectively.

5. Simulations: Irreversible Time Series

So far, we have examined applications of the test to reversible time series, and developed some ways of reducing the effect of dependence among delay vectors. This section will examine the behavior of the test for irreversible time series.

5.1. Chaotic Time Series

Table 4.7 shows the values of $\bar{S}$ and $\sigma(S)$ obtained with $\ell = \tau = 8$ for a number of irreversible time series generated by chaotic maps. The HEN time series consist of the x-component of Hénon's system with the standard parameter values $a = 1.4$ and $b = 0.3$. The LLOG time series obey a linear model with a deterministic noise source, viz. $x_n = 0.5x_{n-1} + \epsilon_n$ where $\{\epsilon_n\}$ is a realization of the chaotic logistic model:

$$\epsilon_n = 4\epsilon_{n-1}(1 - \epsilon_{n-1}) \tag{4.27}$$

(see Figure 3.3 for a phase portrait). Series MODA, MODB and MODC are realizations of the deterministic process

$$X_n = aX_{n-1} \bmod 1, \tag{4.28}$$

where $a = \sqrt{2}$, $a = \sqrt{20}$ and $a = \sqrt{200}$ respectively. The reason for including this model is that for increasing values of a the time series will more and more resemble uniform white noise and the detection of irreversibility is expected to become more difficult.

The chaotic time series for all values of the time series length L considered show values of $\bar{S}$ and $\sigma(S)$ which behave differently than those of the reversible time series examined; the mean $\bar{S}$ approximately grows linearly with L, whereas the standard deviation $\sigma(S)$ increases with L. The value of $\bar{S}$ decreases going from MODA to MODB and MODC as expected. The values for MODC can hardly be distinguished from those of the reversible time series, at least with the current values of the test parameters.

model	$L = 50$ $\bar{S}$	$\sigma(S)$	$L = 100$ $\bar{S}$	$\sigma(S)$	$L = 200$ $\bar{S}$	$\sigma(S)$	$L = 400$ $\bar{S}$	$\sigma(S)$
LOG	3.55	0.41	8.22	0.55	17.57	0.69	36.31	1.16
HEN	3.35	0.41	7.74	0.60	16.51	0.56	34.31	0.82
LLOG	2.24	0.64	5.05	0.86	11.11	1.03	22.80	1.24
MODA	3.77	0.41	8.76	0.52	18.64	0.72	38.41	1.23
MODB	1.34	0.80	3.33	0.60	7.15	0.70	14.58	0.90
MODC	0.31	1.00	0.15	1.01	0.35	1.10	0.22	0.92

Table 4.7: *Mean and standard deviation of S for chaotic time series, estimated from 100 independent realizations.*

As noted above, unlike the cases with reversible time series, the standard error of S increases with L. To obtain some insights into this, we examine the (unconditional) variance of $\widehat{Q}$. Straightforward calculations under the assumption of independence of the reconstruction vectors show that the variance of $\widehat{Q}$ can be expressed as

$$Var\{\widehat{Q}\} = \frac{E(W_{12}^2) + 2(N - T - 1)E(W_{12}W_{23}) + \binom{N-T-1}{2}(E(W_{12}))^2}{\binom{N-T+1}{2}}.$$

(4.29)

The second and third term are zero under the null hypothesis. However, for alternatives, these terms will generally have nonzero values so that the asymptotic variance of S will depend on N as

$$Var\{S\} = 1 + 2(N - T - 1)\frac{E(W_{12}W_{23})}{E(W_{12}^2)} + \binom{N-T-1}{2}\frac{(E(W_{12}))^2}{E(W_{12}^2)}.$$

(4.30)

5.2. Linear non-Gaussian Time Series

model	\$L = 50\$ $\bar{S}$	$\sigma(S)$	\$L = 100\$ $\bar{S}$	$\sigma(S)$	\$L = 200\$ $\bar{S}$	$\sigma(S)$	\$L = 400\$ $\bar{S}$	$\sigma(S)$
NGRP1	0.06	1.03	0.32	1.12	0.55	1.04	1.26	1.04
NGRP2	−0.10	1.05	0.06	0.94	0.29	1.03	0.74	1.22

Table 4.8: *Mean and standard deviation of S for i.i.d. Gaussian distributed time series estimated from 100 simulated independent realizations of non-Gaussian linear processes of order 1 (NGRP1) and order 3 (NGRP2).*

In general, linear non-Gaussian time series are irreversible as shown by Weiss (1975). The NGRP1 time series are obtained with the first order autoregressive model $x_n = 0.3x_{n-1} + \epsilon_n$ where $\{\epsilon_n\} \sim U[-0.5, 0.5]$ and the NGRP2 time series are realizations of the third order autoregressive process $x_n = 0.4x_{n-3} - 0.3x_{n-2} + 0.2x_{n-1} + \epsilon_n$ where $\{\epsilon_n\}$ is a sequence of i.i.d. random variables obtained by squaring independent random variables distributed according to $U[-0.5, 0.5]$. The results for these time series are given in Table 4.8. For NGRP1 and NGRP2, the values of $\bar{S}$ are larger than the expected mean value 0 for reversible time series, which indicates that the test indeed is sensitive to irreversibility due to a non-Gaussian distribution of the noise. Although the values of $\bar{S}$ are positive, which suggest that the test has power against linear non-Gaussian alternatives, the probability of rejection of the null hypothesis for these alternatives is small, and longer time series are required to obtain a power close to 1 for these processes.

5.3. Noisy Irreversible Time Series

model	$L = 50$ $\bar{S}$	$\sigma(S)$	$L = 100$ $\bar{S}$	$\sigma(S)$	$L = 200$ $\bar{S}$	$\sigma(S)$	$L = 400$ $\bar{S}$	$\sigma(S)$
HENN1	0.60	1.01	1.37	1.10	2.99	1.38	6.12	1.45
HENN2	−0.03	1.06	0.20	0.97	0.29	1.04	0.58	1.03

Table 4.9: *Test results for noisy chaotic time series. The average value $\bar{S}$ of S and the standard deviation from the mean, $\sigma(S)$, obtained from 100 independent tests with $L = 50, 100, 200$ for Hénon with 50% observational noise (HENN1) and with 100% observational noise (HENN2).*

In order to get an indication of the sensitivity of the test results with respect to reversible observational noise we contaminated independent Hénon time series with uniformly distributed i.i.d. noise. Table 4.9 shows $\bar{S}$ and $\sigma(S)$ for different time series lengths L. The HENN1 and HENN2 time series are Hénon time series with i.i.d. Gaussian observational noise having a standard deviation of 50% and 100% of the standard deviation of the original time series respectively. The results of the simulations are shown in Table 4.9. The values of $\bar{S}$ are larger than expected under the null hypothesis, and increase with L, indicating that the test has power against this type of alternatives. As expected, the mean S values found are smaller for HENN2 than for HENN1. The small values of $\bar{S}$ obtained for HENN2 suggest that long time series are required in order to obtain a power close to 1.

6. Applications

Stock Return Series   Financial time series such as daily stock prices often exhibit non-stationary behavior. Usually one considers the logarithms of prices which has the advantage that the process generating the time series becomes independent of the actual price levels. The logarithmic prices are, of course non-stationary whenever the price itself is non-stationary, but if the logarithmic data are subsequently transformed into return rates by taking differences,

$$Y_n = \Delta \ln(X_n) = \ln(X_n) - \ln(X_{n-1}), \tag{4.31}$$

stationary time series are usually obtained. As an example, in Figure 4.6 we show a time series of one stock of IBM in the period between 22 October 1996 and 13 February 1998, as well as the logarithmic daily returns in the same

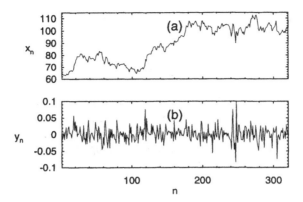

Figure 4.6: *Closing price x_n of IBM stock on the 322 trading days between 22 October 1996 and 13 February 1998 (a) and the first differences $Y_n = \Delta \ln(x_n) = \ln(x_n) - \ln(x_{n-1})$ of the logarithms of the closing prices (b).*

period. Whereas the closing prices are clearly non-stationary, the logarithmic differences appear to be stationary.

It is well known that daily stock return values have little or no autocorrelation. Indeed, if there was autocorrelation it would be easy to exploit this. The absence of autocorrelation, however, does not exclude the possibility that the returns are irreversible. In order to examine this possibility we tested the daily returns of the IBM stock for reversibility. We used a delay of 1 day, embedding dimension $m = 3$, together with $T = \ell = 3$ days and a normalized bandwidth value $d = 0.2$. We obtained the value $S = 1.55$ for our test statistic, which provides no evidence against the hypothesis that the return series is reversible. This result is consistent with a random walk hypothesis for the logarithmic stock value. We have not examined whether the popular GARCH models, in which the variance of the returns depend on the previous variances and realized return values, generally give rise to reversible logarithmic return time series. In that case irreversibility would exclude a GARCH model as the generating mechanism of a time series.

Sunspot Numbers We have applied our test for reversibility to the Wolf sunspot data with $\tau = 1$, $m = 3$, $d = 0.2$, and $T = \ell = 8$. The obtained value for S was 5.78, which provides strong evidence against the hypothesis that the time series is reversible. This confirms the earlier conjecture based on visual inspection of the time series (Figure 4.3) and the phase portrait (Figure

4.4) that a linear Gaussian model, or a linear transformation thereof, is not appropriate for this time series.

7. Alternative Methods to Handle Dependence

Above, we have concentrated on a block method for taking into account dependence among delay vectors. There are several alternative approaches to taking into account dependence, two of which are discussed here. The general problem setting of this section is that of estimating the standard deviation of a U-statistic of dependent observations.

7.1. Estimation of the Covariance Structure

One possible approach consists of estimating the standard deviation of the statistic in a so-called autocorrelation robust way (cf. Robinson and Velasco, 1996). The idea is to take into account correlations among the terms contributing to the test statistic in the estimation of its variance. The unconditional variance of $\widehat{Q}$ is given by

$$
\begin{aligned}
Var\{\widehat{Q}\} &= \frac{1}{\binom{N-T+1}{2}^2} E \left(\sum_{i=1}^{N-T} \sum_{j=i+T}^{N} \sum_{i'=1}^{N-T} \sum_{j'=i'+T}^{N} W_{ij} W_{i'j'} \right) \\
&= \sum_{\substack{k \quad l \\ |k|,|l|,|k-l|<N-T}} N_{kl} C_{kl}
\end{aligned}
\tag{4.32}
$$

where

$$
C_{kl} = E(W_{ij} W_{(i+k)(j+l)}) \tag{4.33}
$$

is the covariance matrix, which is assumed to be independent of i and j for $|i - j| \leq T$, and N_{kl} is the number of terms in the first expression in Equation (4.32) with $i' - i = k$ and $j' - j = l$. Straightforward calculations give $N_{kl} = M(M-1)/2$ with

$$
M = \begin{cases} N - T + 1 - \max(|k|, |l|) & \text{if } kl \geq 0 \\ N - T + 1 - |k| - |l| & \text{if } kl < 0. \end{cases} \tag{4.34}
$$

For mixing time series, it is reasonable to assume that only covariance on small time scales (i.e. C_{kl} for small k and l) will have to be taken into account to obtain a reasonable expression for the variance of $\widehat{Q}$. Upon taking into account the covariance structure up to a time difference s, one obtains

$$
\begin{aligned}
Var\{\widehat{Q}\} &\simeq \frac{1}{\left(\frac{N-T+1}{2}\right)^2} \sum_{k=-s}^{s} \sum_{l=-s}^{s} N_{kl} C_{kl} \\
&\simeq \frac{1}{\left(\frac{N-T+1}{2}\right)} \sum_{k=-s}^{s} \sum_{l=-s}^{s} C_{kl},
\end{aligned}
\tag{4.35}
$$

where the latter approximation is justified for large N by the relation

$$
N_{kl} = \binom{N-T+1}{2} \left\{ 1 + O\left(\frac{|k|+|l|}{N}\right) \right\}.
\tag{4.36}
$$

Equation (4.35) suggests estimation of $Var\{\widehat{Q}\}$ as

$$
\widehat{V}_s = \frac{1}{\left(\frac{N-T+1}{2}\right)} \sum_{k=-s}^{s} \sum_{k=s}^{s} \widehat{C}_{kl},
\tag{4.37}
$$

where $\widehat{C}_{kl}$ is the sample covariance matrix.

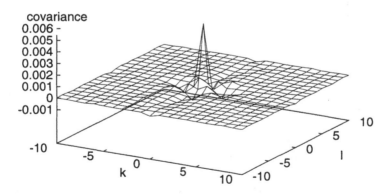

Figure 4.7: *The estimated autocovariance of W_{ij} and $W_{(i+k)(j+l)}$ as a function of k and l for an AR(1) process with $a = 0.95$. The time series has length $L = 400$, and the parameters used are $\tau = 1$, $m = 3$, $d = 0.2$ and $T = 10$.*

Figure 4.7 shows the estimated covariance structure $\widehat{C}_{kl}$ of W_{ij} for a realization of an AR(1) process with parameter $a = 0.95$. As expected, the estimated

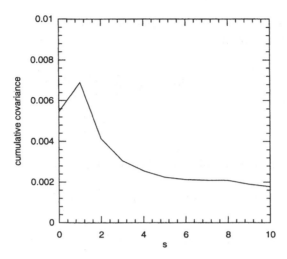

Figure 4.8: *Cumulative covariance corresponding to Figure 4.7.*

covariance $\widehat{C}_{kl}$ is largest for small values of k and l. The covariance decays to zero quickly in most directions, although there appears to be a relatively slowly decaying negative covariance along the lines $k = 0$ and $l = 0$. Figure 4.8 shows the estimated cumulative covariance

$$\sum_{k=-s}^{s} \sum_{l=-s}^{s} \widehat{C}_{kl}$$

for the same time series. The cumulative covariance reaches a plateau for $s > 5$, so that it can be expected that taking into account covariance up to $|k| = 5$ and $|l| = 5$ will give a reasonable estimate of the variance of $\widehat{Q}$. For $s = 5$, the cumulative covariance is 2.249×10^{-3} which, upon substitution in Equation (4.37) with $N = 398$ and $T = 10$, gives the estimate $(\widehat{V}_5)^{1/2} = 1.726 \times 10^{-4}$ for the standard deviation of $\widehat{Q}$. We found $\widehat{Q} = 2.290 \times 10^{-6}$, which gives $\widehat{Q}/(\widehat{V}_5)^{1/2} = 0.01327$. For this time series, the block method (with $\ell = 10$) gave the very similar estimate $\sigma_c(\widehat{Q}) = 1.721 \times 10^{-4}$, corresponding to an S-value of 0.01330.

The method just described requires computation times which are prohibitive for the estimation of true p-values based on a large number of repeated calculations, as done for the block method. Numerical results suggest that the method is reliable (and possibly preferable over the block method) for longer time series (with $L > 400$ say). Also, we observed that larger values of s typically give estimates $\widehat{V}_s$ with reduced bias but larger variance.

7.2. Block Randomization

Another method for taking into account dependence is based on exploiting the fact that the contributions from blocks of length ℓ can, for all practical purposes, be considered as independent if the block length ℓ is sufficiently large. This knowledge was also used in the calculations of the conditional variance of $\widehat{Q}$ by the block method. Rather than determining only the variance of $\widehat{Q}$, we can go one step further and construct a reference distribution of $\widehat{Q}$ under the null hypothesis. The idea is to randomize the data by applying P at random (with probability $1/2$) to the blocks of reconstruction vectors. That is, given a block, P is applied with probability $1/2$ to either all or none of the reconstruction vectors in the block. This approach would roughly require ℓ to be sufficiently large to ensure that the cross-covariance C_{kl} is negligibly small compared to $E(W_{ij}^2)$ whenever either k or l is larger than ℓ.

This allows for the calculation of a p-value, that is, the relative proportion of the reference distribution for which the randomized values of $\widehat{Q}$ exceed the observed value of $\widehat{Q}$. For a test with size α one rejects the null hypothesis whenever the p-value is smaller than α. For the time series discussed in the previous section (AR(1) with $a = 0.95$) using $\ell = 10$ we found, again with $\tau = 1$, $m = 3$ and $T = 10$, a p-value of 0.45 based on 1000 randomizations. This implies that the observed value of $\widehat{Q}$ was exceeded by 45% of the randomized values, suggesting that there is no reason to reject the null hypothesis.

We note that this randomization approach differs from bootstrap methods like the block bootstrap proposed by Künsch (1989). Our aim here is not to generate bootstrap time series for constructing confidence intervals, but rather to generalize a randomization method (consisting of randomly applying the map P to individual vectors with probability $1/2$) which is justified for independent vectors, to delay vectors. This approach is in fact more closely related to Monte Carlo testing than to bootstrap methods.

8. Bandwidth

In the applications to simulated time series the embedding parameters and the bandwidth were fixed. A thorough investigation into the size and power of the test for various embeddings and bandwidth parameters is beyond the scope of this book, but a few remarks will be made on the bandwidth here.

The optimal bandwidth, as far as the power of the test is concerned, will of course depend on the underlying generating process of the time series. For each irreversible process there will be optimal settings of the parameters m and d, which in general are unknown. For very small bandwidths the power of the test will be small because the test uses only local information on the measure μ. In that case a delay vector $\boldsymbol{X}_i$ has effectively only a small number of neighbors $\boldsymbol{X}_j$ that contribute to the sum of W_{ij} over j in the test statistic.

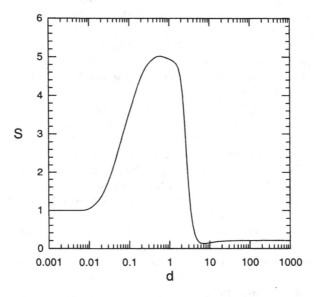

Figure 4.9: S as a function of the bandwidth d for a Hénon time series of length $L = 100$, with test parameters $m = 3$, $\tau = 1$ and $\ell = T = 10$.

The problem is analogous to that arising in estimation of correlation integrals at small values of the radius r. Because small distances become less frequent for larger embedding dimensions m, this effect is more severe for larger values of m. In the limit $d \to 0$, only the contribution W_{ij} corresponding to the smallest inter-point distance makes a dominant contribution to $\widehat{Q}$. For $\ell = 1$, it can be readily checked that S converges to either 1 or -1 for $d \to 0$, depending on whether the particular W_{ij} which contains the smallest distance (among $\|\boldsymbol{X}_i - \boldsymbol{X}_j\|$ and $\|\boldsymbol{X}_i - P\boldsymbol{X}_j\|$) is a distance between two time-forward delay vectors or a time-forward and a time-backward delay vector (i.e. whether this dominant term W_{ij} is positive or negative for $d \to 0$). For block lengths $\ell > 1$ the situation is similar, as there will be one term W_{ij} which is dominant for $d \to 0$ (the block which contains the smallest distance) and depending on whether this term is positive or negative, S will converge to either 1 or -1. An example of this phenomenon is shown in Figure 4.9 for a Hénon time series of length $L = 100$. Clearly, the test does not have power against any alternative in this limit.

On the other hand, for very large bandwidths there is *a priori* no reason why the test should have no power. In the limit $d \to \infty$, both $\widehat{Q}$ and $\sigma_c(\widehat{Q})$ approach 0, but $S = \widehat{Q}/\sigma_c(Q)$ has a well-defined limit. From Equations (4.11) and (4.12) one finds, up to leading order in $1/d$,

$$W_{ij} \sim \frac{1}{2d^2} \left(\|X_i - PX_j\|^2 - \|X_i - X_j\|^2 \right), \qquad (4.38)$$

which gives

$$\lim_{d \to \infty} S = \frac{\sum_{i,j}' \|X_i - PX_j\|^2 - \|X_i - X_j\|^2}{\left(\sum_{i,j}' \left(\|X_i - PX_j\|^2 - \|X_i - X_j\|^2 \right)^2 \right)^{\frac{1}{2}}}. \qquad (4.39)$$

Indeed, Figure 4.9 shows a value of S that converges to a finite value when the bandwidth tends to infinity. Recall that the test for finite bandwidth values has power against any fixed alternative. Whether it still has this property in the limit where d tends to infinity is questionable, since positive definite-ness of the expected value of the test statistic was proved only for finite positive values of the bandwidth.

To obtain some insights into the dependence of the optimal bandwidth on the length of the time series, we adopt the following convention: given an irreversible stochastic process and embedding parameters m and τ, the value d^* of the bandwidth d for which the expected value of S is largest is considered the optimal bandwidth. It may appear natural, at first, that the optimal bandwidth depends on the length of the time series. In nonparametric function estimation with kernel methods, for example, the optimal bandwidth is known to decrease with the sample size. Inspection of the test statistic, S, however, shows that the optimal bandwidth for the test is asymptotically independent of L. Taking into consideration all inter-point distances, for large values of the time series length L, the expected value of $\widehat{Q}$ is Q and the expected value of $\sigma(S)$ behaves as L^{-1} for any fixed value of the bandwidth d. Therefore, doubling the number of delay vectors will result in a doubling of the expected value of S for all values of d. The bandwidth d^* for which S attains its maximum thus remains unchanged. Figure 4.10 shows S as a function of d for Hénon time series of length $L = 70, 100, 140, 200$ and the value of d^* practically does not change with L.

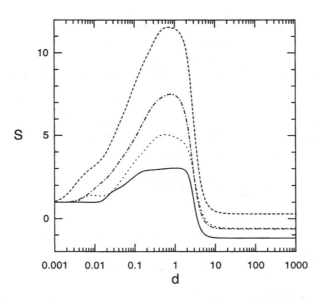

Figure 4.10: *S as a function of the bandwidth d for a Hénon time series of length*
$L = 70, 100, 140, 200$, *with test parameters* $m = 3$, $\tau = 1$ *and* $\ell = T = 10$.

The fact that the optimal bandwidth in nonparametric function estimation decreases with the sample size is related to a bias/variance trade-off; for
smaller bandwidths the bias decreases but the variance increases. For increasing sample sizes the variance decreases and the optimal bandwidth becomes
smaller. Here the bias plays no role, as our statistic is unbiased for all choices
of the bandwidth. The optimal bandwidth here is determined by the optimal
ratio of $\widehat{Q}$ and its standard deviation under the null hypothesis $\sigma_c(\widehat{Q})$, and
turns out to be asymptotically independent of the sample size.

   The problem remains as to how the bandwidth should be chosen in practice,
where one usually has only one observed time series. A straightforward way
to approach this problem is by using the fact that the optimal bandwidth d^*
for a given irreversible process is independent of the time series length. One
could thus use a small initial segment of the time series, calculate S in a range
of d-values, and take the value of d for which this function has a maximum
as an estimate of the optimal bandwidth $d*$. Then the test can be applied to
the remaining part of the time series, with a bandwidth equal to the estimate
of d^*. In fact this approach is closely related to the idea of cross-validation in
statistics.

Alternatively, one can avoid the problem of the selection of a value for the bandwidth parameter by determining the ratio S for a number of bandwidths and then taking the supremum S^* of S over the bandwidth d,

$$S^* = \sup_d S(d), \qquad (4.40)$$

as the statistic of interest. A similar approach was suggested by Le Cessie and Van Houwelingen (1993) for a goodness of fit test for logistic regression models. The null distribution of S^* can be found numerically using the a block randomization method described in section 7.2 of this chapter.

9. Concluding Remarks

In this chapter, the dynamical properties of linear Gaussian random processes and transformations of these have been related to reversibility. We have proposed a test for the null hypothesis that a time series is reversible. The motivation being that a rejection of the null hypothesis (the detection of irreversibility) allows exclusion of a linear Gaussian random process as a candidate model for the generating mechanism of a time series. Because reversibility is preserved by a number of transformations, a larger class of candidate processes can be excluded. For example, all static, not necessarily monotonous, transformations of linear Gaussian random processes are contained within this class. The test differs in a fundamental way from existing tests for linearity which also allow for static transformations because the static transformations need not be one-to-one.

The test statistic is of a conventional type in that it does not require the generation of surrogate data to obtain its mean and variance under the null hypothesis, which can be calculated exactly for independent delay vectors. We applied the method to different model time series. For reversible processes, including static transformations of several linear Gaussian random processes, no significant deviations of the test statistic from the values expected under the null hypothesis were found, using a Theiler correction to suppress bias, and a block-based method for the calculation of the variance of the test statistic.

The power of the test in applications to irreversible time series depends strongly on the type of data. To examine the limitations of the test, we have performed a number of numerical simulations for relatively short time series. For irreversible time series, the average S-values were found to increase with the length of the time series. The numerical results for deterministic, irreversible time series contaminated with i.i.d. (reversible) observational noise suggest that

the method can detect irreversibility in the presence of fairly large amounts of
i.i.d. observational noise.

Recently, our test for reversibility was applied to ElectroEncephaloGrams
(EEGs) as a means of distinguishing between normal and epileptiform EEG
(Van der Heijden et al., 1996). Normal EEGs were found to be reversible,
whereas EEGs of mesial temporal lobe epilepsy are irreversible. Although
the clinical relevance of these results is limited because EEGs clinically can be
classified well by visual inspection, the results are important from a theoretical
point of view because they suggest that the mechanisms responsible for the
generation of normal and epileptiform EEGs are fundamentally different. The
EEGs were linearly filtered by the measurement apparatus before registration.
If the original signal was a static transformation of a linear Gaussian random
process, it could have become irreversible due to this filtering. Therefore, we
can not exclude a static transformation of a linear Gaussian random process as
a possible explanation for the observed epileptiform EEGs. However, a linear
Gaussian process can still be ruled out, as such a signal should have remained
reversible under linear filtering.

Recently, Hoekstra et al. (1997) applied the test for reversibility in a study
of the effect of the class Ic drug cibenzoline on the dynamics of sustained
atrial fibrillation during the pharmacological conversion of atrial fibrillation in
goats. Prior to nonlinear analysis, the electrograms were tested for reversibility
in order to avoid possible spurious results due to data generated by linear
Gaussian random processes. The electrograms corresponded to various types
of atrial fibrillation, and it was found that the values of $\hat{Q}$ are related to the
type of atrial fibrillation.

Konings et al. (1994) clinically classified human electrograms using a mapping system for visualizing the space-time data, and by taking the average
number of wavelets wandering in a given region of the atrium as a measure of
complexity. In fact the reversibility test statistic $\hat{Q}$ is negatively correlated with
this measure of complexity, suggesting that irreversibility is less pronounced
in more complex types of atrial fibrillation. A heuristic explanation for this is
that more complex types of atrial fibrillation resemble i.i.d. (reversible) noise
more closely.

CHAPTER 5

DETECTING DIFFERENCES BETWEEN RECONSTRUCTION MEASURES

In this chapter we propose a test for the null hypothesis that two multivariate samples are drawn from the same multi-dimensional probability measure. The application to reconstruction measures provides a test for the null hypothesis that two time series are independent realizations of the same stationary process.

1. Introduction

In time series analysis, the question whether two sets of delay vectors are drawn from the same underlying multi-dimensional probability measure is relevant in many situations encountered. For example, if a model time series is to be compared with the original time series, objective criteria for the quality of the model are desirable, if not necessary. Also, in examining the stationarity of a time series it is important to be able to detect slow changes in its behavior. These changes may not be manifest in the linear properties of the time series, usually determined with power spectra or estimated parameters of linear models. Also, in the study of multiple time series, from a spatio-temporal dynamical system, for example, which may be non-homogeneous in space, a preliminary analysis usually consists of comparing time series measured at different sites. With these applications in mind, it is the aim of this chapter to propose a test for the hypothesis that two time series have identical reconstruction measures, a necessary condition for two stationary time series that are independent generalizations of the same process.

No assumptions are made regarding the model underlying the time series. The only assumption made is that the two time series are segments of mixing stationary time series. The two sets of reconstruction vectors can be then considered samples from the reconstruction measures μ_1 and μ_2. The null hypothesis is

$$H_0: \quad \mu_1 = \mu_2 \tag{5.1}$$

81

The characterization of time series by their distributions of delay vectors, or equivalently, by their reconstruction measures, has become standard in the nonlinear time series literature (Grassberger *et al.*, 1991). In this general context typically nonparametric methods are called for. Several nonparametric methods for comparing time series that do not focus on a specific class of time series models but rather on delay vector measures (or reconstruction measures) can be found in the literature. Wright and Schult (1993) proposed a method for identifying the presence of a time series with known structure within an observed time series. Albano *et al.* (1995) compare delay vector distributions of a pair of time series by applying a Kolmogorov-Smirnov test to the two correlation integrals. Kantz (1994) introduced a method for comparing time series using correlation integrals and cross-correlation integrals. The remark that one of the quantities Kantz considers under certain conditions and in a certain limit can be interpreted as a distance between two delay vector distributions, has motivated the test statistic used here.

The test described in this chapter is based on a distance notion between two multi-dimensional probability measures similar to that used in the previous chapter. An unbiased consistent estimator of the square of this distance is constructed, and its variance calculated conditionally on the set of observed reconstruction vectors. The test is initially developed under the assumption of independence of all vectors. That is, a general nonparametric test is constructed for the null hypothesis that two samples of independent vectors are drawn from the same probability measure. Later it is examined how the assumption of independence can be relaxed so that the test can be applied to time series.

2. A Distance between Probability Measures

distance between measures

In this section a distance between two probability measures, μ_1 and μ_2 in $\mathbf{R}^m$ is defined. As for the reversibility test, the square of this distance can be estimated in an unbiased manner from two sets of vectors sampled independently from μ_1 and μ_2. We consider the case of two samples, $\{\mathbf{X}_i\}_{i=1}^{N_1}$ and $\{\mathbf{Y}_i\}_{i=1}^{N_2}$ from μ_1 and μ_2 respectively. The quantity Q is defined as

$$Q = (\mu_1 - \mu_2, \mu_1 - \mu_2), \tag{5.2}$$

where the bilinear form is given by Equation (4.4). As stated in Proposition 4.1, $Q = 0$ if μ_1 and μ_2 are equal, and $Q > 0$ otherwise.

To construct an unbiased estimator $\widehat{Q}$ for Q, it is convenient to rewrite Equation (5.2) as

$$Q = Q_{11} + Q_{22} - 2Q_{12}, \tag{5.3}$$

where

$$Q_{kl} = \iint h(s,t)\,\mathrm{d}\mu_k(s)\,\mathrm{d}\mu_l(t), \qquad k,l \in 1,2 \tag{5.4}$$

with

$$h(s,t) = \mathrm{e}^{-\|s-t\|^2/(4d^2)}. \tag{5.5}$$

From Equations (5.3) and (5.4) it follows that

$$\widehat{Q} = \widehat{Q}_{11} + \widehat{Q}_{22} - 2\widehat{Q}_{12}, \tag{5.6}$$

with

$$
\begin{aligned}
\widehat{Q}_{11} &= \frac{1}{\binom{N_1}{2}} \sum_{1 \le i < j \le N_1} h(X_i, X_j) \\
\widehat{Q}_{22} &= \frac{1}{\binom{N_2}{2}} \sum_{1 \le i < j \le N_2} h(Y_i, Y_j) \\
\widehat{Q}_{12} &= \frac{2}{N_1 N_2} \sum_{i=1}^{N_1} \sum_{j=1}^{N_2} h(X_i, Y_j)
\end{aligned}
\tag{5.7}
$$

is an unbiased U-estimator of Q.

The variance $Var_c\{\widehat{Q}\}$ of $\widehat{Q}$ under the null hypothesis and conditionally on the set of $N_1 + N_2$ observed vectors is a function of N_1, N_2 and the set of $N_1 + N_2$ vectors. If we define $N = N_1 + N_2$ and

$$z_i = \begin{cases} X_i & \text{for } 1 \le i \le N_1 \\ Y_{i-N_1} & \text{for } N_1 < i \le N, \end{cases} \tag{5.8}$$

the variance under the null hypothesis conditionally on $\{z_i\}_{i=1}^N$ is

$$Var_c\{\widehat{Q}\} = \frac{2(N-1)^2(N-2)}{N_1(N_1-1)N_2(N_2-1)(N-3)} \left(\frac{1}{\binom{N}{2}} \sum_{1 \le i < j \le N} \psi_{ij}^2 \right), \tag{5.9}$$

in which

$$\psi_{ij} = H_{ij} - g_i - g_j, \tag{5.10}$$

where

$$H_{ij} = h(z_i, z_j) - \frac{1}{\binom{N}{2}} \sum_{1 \le i' < j' \le N} \sum h(z_{i'}, z_{j'}),$$ (5.11)

and

$$g_i = \frac{1}{N-2} \sum_{\substack{j \\ j \ne i}} H_{ij}.$$ (5.12)

A derivation of this result is given in Appendix B, which also gives an expression for the unconditional variance. In fact, the latter is more easily derived, but conditioning on the set of vectors has the advantage that an exact expression for the variance is obtained. The statistic S defined as the ratio

$$S = \frac{\widehat{Q}}{\sqrt{Var_c\{\widehat{Q}\}}}$$ (5.13)

has zero mean and standard deviation equal to one under the null hypothesis, and under the assumption that all vectors are independent. The next section discusses adaptations of the test for dependent delay vectors.

3. Application to Reconstruction Measures

For two time series $\{X_i\}_{i=1}^{L_1}$ and $\{Y_i\}_{i=1}^{L_2}$ the sets of m-dimensional reconstruction vectors with delay τ consists of $N_1 = L_1 - (m-1)\tau$ and $N_2 = L_2 - (m-1)\tau$ vectors respectively. The major problem for the application of virtually all statistical methods in the field of time series analysis is the presence of dependence among reconstruction vectors. The reconstruction vectors are dependent even if both time series consist of independent, identically distributed (i.i.d.) observations. The mere construction of the sets of reconstruction vectors from the two time series introduces dependence among the reconstruction vectors, as they have elements in common. Dynamical structure in the time series will introduce additional dependence.

Using examples, we will first examine the properties of the test if dependence is ignored. A modification of the test similar to that made for the reversibility test described in the previous chapter then is shown to work well for many simulated time series. Before each analysis both time series are rescaled by the same factor so that the overall sample standard deviation of the $N = N_1 + N_2$ observations equals one. The parameters for the construction of the delay vectors are fixed at $m = 3$ and $\tau = 1$ and the bandwidth parameter at $d = 0.2$. The mean and standard deviation of S are estimated

process	$\ell = T = 1$ $\bar{S}$	$\sigma(S)$	$\ell = T = 5$ $\bar{S}$	$\sigma(S)$	$\ell = T = 10$ $\bar{S}$	$\sigma(S)$	$\ell = T = 15$ $\bar{S}$	$\sigma(S)$
UNO	−0.03	1.05	0.02	0.99	0.00	0.99	0.02	0.99
LOG	0.08	1.99	−0.09	1.38	0.15	1.28	−0.15	0.98
HEN	−0.38	2.39	−0.05	1.43	−0.10	1.34	−0.12	1.02

Table 5.1: *Mean value $\bar{S}$ and standard deviation $\sigma(S)$ of S estimated from 100 independent realizations of uniform i.i.d. random data (UNO), logistic map data (LOG) and Hénon map data (HEN). All time series have a length L of 200.*

numerically by repeated applications of the test to pairs of independent realizations of a given process. For each process, 100 independent values of S are determined and from the results, the mean $\bar{S}$ and standard deviation $\sigma(S)$ of S are estimated. The three processes we consider are uniformly distributed random data, logistic map data and Hénon map data. All time series have a length of $L = 200$, so that with $m = 3$ and $\tau = 1$ the two sets consist of $N_1 = N_2 = L - (m - 1)\tau = 198$ reconstruction vectors.

The results are given in Table 5.1. As in the previous chapter, the parameters T and ℓ in this table refer to the Theiler correction and the block length respectively. The effects of the Theiler correction and the block method are discussed below. First we concentrate on the un-adapted method ($\ell = T = 1$) the results of which are given in the first column of Table 5.1. For the un-adapted method only the uniformly distributed i.i.d. time series gives estimates of $\bar{S}$ and $\sigma(S)$ that are close to 0 and 1 respectively. For comparison, if the vectors were independent, the expected value of $\bar{S}$, which is the mean value of S in the 100 calculations, would be 0 and the standard error of $\bar{S}$ would be 0.1. The expected value of $\sigma(S)$ would be 1 with a standard error of about $0.1/\sqrt{(2)} \approx 0.07$, the value one would obtain if S would have a standard normal distribution. For the logistic map the estimate of $\bar{S}$ for $\ell = T = 1$ is close to zero, but the standard deviation of S is nearly twice as large as the nominal value 1, which suggests that a test with $\ell = T = 1$ is anti-conservative (i.e. has a larger than nominal type I error). For the Hénon time series, the standard deviation of S is also too large, and S furthermore shows a negative bias. From these results we conclude that the dependence introduced by the construction of delay vectors appears to be of minor influence on the null distribution, but dynamical dependence, as suggested by the results for the two chaotic maps, should not be ignored.

The dependence among delay vectors was investigated in the previous chapter for the reversibility test where it was shown that the bias in the test statistic can be reduced by leaving out the contributions of (i, j) pairs of indices for which $|i - j|$ is smaller than T, where T is the Theiler correction. Note that a Theiler correction here involves only $\widehat{Q}_{11}$ and $\widehat{Q}_{22}$, because $\widehat{Q}_{12}$ contains contributions involving pairs of delay vectors from different time series, which are assumed to be independent.

A Theiler correction reduces the bias but does not allow the use of the expression for the variance of the estimator as derived for independent vectors. Strongly dependent delay vectors will affect the variance of the estimator, regardless of whether a bias has been removed or not. Even if (i, j) pairs which have $|i - j| \geq T$ are pairwise independent, the contributions from (i, j) and $(i + k, j + l)$ for k, l small can be correlated. Again this can be corrected for using a block method.

The Block Method   As was done for the reversibility test, the (i, j)-plane of indices is divided in squares of size $\ell \times \ell$ and the average values

$$h'(i', j') = \frac{1}{\ell^2} \sum_{p=1}^{\ell} \sum_{q=1}^{\ell} h(z_{i'\ell+p}, z_{j'\ell+q}), \tag{5.14}$$

in each of these squares are used in place of $h(i, j)$. Again, using h' instead of h in the definition of $\widehat{Q}$ in Equation (5.6), the conditional variance in terms of h' is analogous to Equation (5.9) with h' substituted for h. For ℓ large enough, non-overlapping blocks of length ℓ within the multi-dimensional series of delay vectors can be considered to be independent, and in this way one can take into account short-range (on a time scale smaller than ℓ) dependence among delay vectors. Because it is to be expected that ℓ, like T, should be chosen larger than the time scale on which dependence decays we have performed our calculations with $\ell = T$. The results in Table 5.1 clearly show two effects of the block method when compared with the naive method ($\ell = T = 1$). As a result of the Theiler correction the bias is reduced to a value near zero. The block method leads to improved estimates of the standard deviation of $\widehat{Q}$ as shown by the fact that the values of the standard deviations in S are much closer to 1 for the block method than for the naive method.

The kernel function used is positive definite, which calls for a one-sided rejection criterion, rejecting when the estimated value of S is too large. Consider rejecting the null hypothesis if the value of S is larger than 3. If S indeed has mean zero, and unit standard deviation, and assuming furthermore that S has a unimodal distribution, the probability of finding a value of S larger than 3 is smaller than 0.05 (Pukelsheim, 1994). Since the latter inequality applies to the two-sided test case, this inequality will be conservative when used in the one-sided case. In additional numerical simulations with the block method the

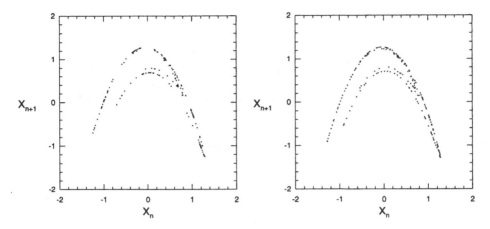

Figure 5.1: *Phase portraits of two time series of length* 200 *generated with Hénon's model with* $a = 1.35$, $b = 0.31$ *(left) and the standard parameter values* $a = 1.4$, $b = 0.3$ *(right).*

numbers of S-values larger than or equal to 3 were 12, 22 and 11 out of 10^3 for the uniform random data, for the logistic map data and the Hénon map data respectively. These values confirm that the probability of rejecting the null hypothesis, using $S \geq 3$ as a rejection criterion is well below 0.05 under the null hypothesis.

Example As an example of an alternative, that is, a pair of processes for which the null hypothesis does not hold, we compare the reconstruction measures of pairs of time series generated with Hénon's model, $X_{n+1} = 1 - aX_n^2 + bX_{n-1}$, for slightly different model parameters. For each first time series the parameter values $a = 1.35, b = 0.31$ were used, while for each second series the standard parameter values $a = 1.4, b = 0.3$ were used (cf. Kantz, 1994). The phase portraits of typical examples of two time series that are compared here is shown in Figure 5.1. The results found for $d = 0.2$ with the block method ($\ell = T = 15$) are $\bar{S} = 2.68$ and $\sigma(S) = 1.39$. Using the rejection criterion $S > 3.0$, there were 355 rejections of the null hypothesis out of the 10^3 simulations. Actually, this choice for the bandwidth is not optimal. For $d = 0.1$ we found $\bar{S} = 4.25$, $\sigma(S) = 1.22$ and there were 984 rejections out of 10^3 simulations. In practice of course the optimal value of the bandwidth is generally not known.

4. Bandwidth

This section briefly discusses the choice of the bandwidth parameter d which sets the length scale of the smoothing and hence determines the length scale at which the two reconstruction measures are compared. By choosing the bandwidth relatively small, $\widehat{Q}$ will pick up local differences in the reconstruction measures. Taking the bandwidth too small, however, will give rise to poor statistics as can be seen from the behavior of $\widehat{Q}$ in the limit $d \to 0$. In this limit, the sums in Equation (5.6) are dominated by one term only: the term corresponding to the smallest distance in the set of vectors. For large d, the reconstruction measures are compared on a huge length scale and it can be expected that small scale differences become practically indistinguishable. We expect to find an optimal value for the bandwidth at the trade-off of these two effects. It is known that the optimal bandwidth for density estimators depends on the number of observations and decreases for larger numbers of data. However, the problem we address here is not that of optimally estimating densities, but rather that of finding the bandwidth for which the test is most powerful. As for the reversibility test the optimal bandwidth is asymptotically independent of the time series length.

First we consider the bandwidth dependence of the three different terms in the estimator $\widehat{Q}$ defined by Equation (5.6). The first term contains contributions from distances between pairs of delay vectors of the first time series only. The second term is a function of the second time series only, while the third term contains the cross terms which involve both time series. For two time series generated with Hénon's model at different parameter values (see the example in the previous section), we calculated (with the block method, $\ell = T = 15$) the separate contributions $\widehat{Q}_{kl}$ as a function of d.

Figure 5.2 shows a log–log plot of the estimates versus the bandwidth parameter d. For large d the curves of the three estimators practically coincide while for small d the curves diverge from one another. There is a qualitative correspondence between the behavior of the curves in Figure 5.2 and of the estimators of their square Θ-kernel analogs: the traditional correlation integrals

$$C_{kk}(d) = \iint \Theta(d - \|s - t\|) \, d\mu_k(s) \, d\mu_k(t) \qquad \text{for } k \in \{1, 2\}, \qquad (5.15)$$

and the cross-correlation integral (Kantz, 1994),

$$C_{12}(d) = \iint \Theta(d - \|s - t\|) \, d\mu_1(s) \, d\mu_2(t). \qquad (5.16)$$

The scaling behavior of correlation integrals of deterministic time series is also present for Q_{11} and Q_{22}. It can be proved that if μ_k has correlation dimension

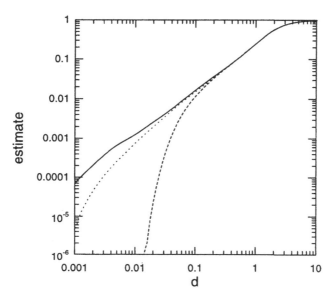

Figure 5.2: *Estimated values (dimensionless) of Q_{11} (solid line), Q_{22} (dotted line) and Q_{12} (dashed line) as a function of the bandwidth parameter d (dimensionless) for two time series of length $L = 200$, generated by slightly different Hénon models.*

D_2, i.e.

$$C_{kk} \sim d^{D_2} \qquad \text{for } d \to 0, \tag{5.17}$$

then Q_{kk} obeys

$$Q_{kk} \sim d^{D_2} \qquad \text{for } d \to 0. \tag{5.18}$$

Next the dependence of S on the bandwidth is considered. Figure 5.3 shows S as a function of d for the pair of time series used for Figure 5.2. The function clearly has a pronounced maximum value near $d \approx 0.013$. We will show that the bandwidth value for which S attains its maximum is asymptotically independent of the time series lengths N_1 and N_2, for a fixed ratio N_1/N. To examine the N-dependence of the optimal value of the bandwidth parameter d for detecting differences between two probability measures we consider a situation where the two sets of reconstruction vectors are samples from two different reconstruction measures μ_1 and μ_2, i.e. we consider a fixed alternative. In general, for a fixed value of the bandwidth d, $E(S)$ will depend on

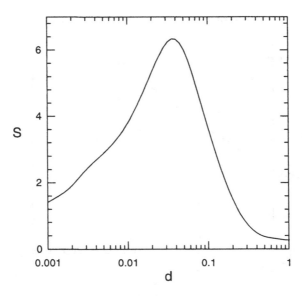

Figure 5.3: *Number of standard deviations S (dimensionless) as a function of the bandwidth d (dimensionless) corresponding to Figure 5.2.*

$p = N_1/N$ and N as

$$E(S) \sim p(1 - p)N \qquad \text{for } N \text{ large,} \qquad (5.19)$$

as a result of the asymptotic (large N) behavior of $\widehat{Q}$ and $Var_c\{\widehat{Q}\}$, which are the numerator and the denominator in the expression for S given in Equation (5.13) respectively. The estimator $\widehat{Q}$ given in Equation (5.6) for large N becomes sharply distributed around the true value Q which depends on μ_1, μ_2 and d, and by Equation (5.9) we have

$$1/\sqrt{Var_c\{\widehat{Q}\}} \sim p(1 - p)N \qquad \text{for } N \text{ large.} \qquad (5.20)$$

Taking into account the dependence on d, one has

$$E(S) \simeq p(1 - p)Ng_{\mu_1,\mu_2}(d) \qquad \text{for } N \text{ large,} \qquad (5.21)$$

where $g_{\mu_1,\mu_2}(d)$ is some function of d which depends on μ_1 and μ_2. Thus for large N and given μ_1, μ_2 and $p = N_1/N$, the value of d for which $E(S)$ has its maximum is asymptotically independent of N.

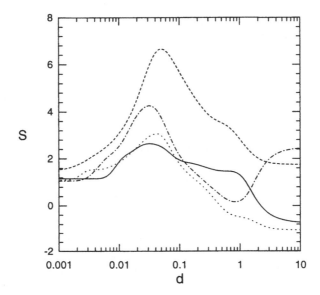

Figure 5.4: *Number of standard deviations S (dimensionless) as a function of the bandwidth d (dimensionless), each for a pair of two Hénon time series with slightly different parameters, and with $N_1 = N_2 = 70$ (solid line), $N_1 = N_2 = 100$ (dotted line), $N_1 = N_2 = 140$ (dash-dotted line) and $N_1 = N_2 = 200$ (dashed line).*

This result is confirmed by numerical experiments in which we determined the value of S as a function of the bandwidth for pairs of time series with increasing lengths N_1 and N_2 with $N_1/N = 1/2$. Independent realizations of pairs of Hénon time series with different parameter values were used. The results for $N_1 = 70, 100, 140, 200$ are shown in Figure 5.4. The bandwidth for which S attains its maximum is practically independent of the time series length.

This result justifies the following approach to choosing the bandwidth. The function $g_{\mu_1,\mu_2}(d)$ appearing in Equation (5.21) can be estimated using a relatively small number of the available vectors, for example those obtained from two small initial segments of both time series. This estimated function of d can be used together with Equation (5.21) as a guide for the estimation of d^*, the value of d for which the method is most powerful given the type of data under consideration. The last segments of both time series can then be used to obtain an independent value of S using the estimated optimal bandwidth.

Alternatively, as argued in Chapter 4, choosing a bandwidth value can be avoided by using an overall test statistic such as

$$S^* = \sup_d S(d). \tag{5.22}$$

For the example function $S(d)$ shown in Figure 5.2 the value of S^* is about 6.4. It can however not be concluded that the curve differs from the expected curve by 6.4 standard deviations. By considering only the optimum value, one is selecting the largest value in a realization of a random function, and the probability of rejecting the null hypothesis while it holds (the type I error) would become larger than for the fixed bandwidth test. One way of dealing with this problem is to develop a theory concerning the random function S of the bandwidth d under the null hypothesis, and derive the distribution of its maximum S^* under the null hypothesis. However, the null distribution of S^* could alternatively be determined using a conditional Monte Carlo procedure. The block method used for calculating the conditional standard deviation can be extended straightforwardly to a Monte Carlo method based on block randomization. Because under the null hypothesis all blocks can be considered to be independent, the observed set of reconstruction vectors can be randomized by arbitrarily redistributing the blocks of observed reconstruction vectors (of length ℓ) over two sets, keeping the numbers N_1 and N_2 fixed. Repeating this procedure and calculating S^* for each randomized set of vectors, one can construct the conditional distribution of S^* under the null hypothesis. By determining the relative frequency with which the test statistic calculated for the randomized data exceeds that of the original data, a one-sided p-value is obtained.

Above, the bandwidth dependence of the estimators $\widehat{Q}$ and S was discussed in the context of the optimal choice for statistical inference. However, the bandwidth dependence of Q_{11}, Q_{22} and Q_{12} may contain some additional relevant information. For example, an informative statistic may be the largest bandwidth above which the three estimates of the Gaussian kernel correlation integrals $\widehat{Q}_{11}$, $\widehat{Q}_{22}$ are the same as $\widehat{Q}_{12}$ within a given accuracy ϵ (cf. Kantz, 1994). This statistic, denoted by d_ϵ, is defined as the minimum bandwidth above which both $|\log \widehat{Q}_{11} - \log \widehat{Q}_{12}| < \epsilon$ and $|\log \widehat{Q}_{22} - \log \widehat{Q}_{12}| < \epsilon$. In some cases the bandwidth d_ϵ allows a clear physical interpretation. For example, when a noise-reduced time series is compared to the original time series, d_ϵ will be a rough estimate of the amplitude of the noise in the original signal.

4.1. Bandwidth Limits

As for the test statistic considered in the previous chapter, S converges to finite limits as $d \to 0$ and $d \to \infty$. We define $N = N_1 + N_2$ and consider

$$z_i = \begin{cases} X_i & \text{for } 1 \le i \le N_1 \\ Y_{i-N_1} & \text{for } N_1 < i \le N \end{cases} \tag{5.23}$$

as given vectors in $\mathbf{R}^m$. Given the set of vectors $\{z_i\}_{i=1}^N$, we consider random divisions of their indices into two groups of sizes N_1 and N_2. We obtain

$$\widehat{Q} = \frac{1}{\binom{N_1}{2}} \sum_{\substack{i,j \in D \\ i<j}} h(z_i, z_j) + \frac{1}{\binom{N_2}{2}} \sum_{\substack{i,j \in D^c \\ i<j}} h(z_i, z_j)$$
$$- \frac{2}{N_1 N_2} \sum_{i \in D} \sum_{j \in D^c} h(z_i, z_j), \tag{5.24}$$

where D is a set of N_1 indices randomly selected without replacement from $\{1, 2, \ldots, N\}$ and D^c is the complementary set of indices. For $d \to 0$, the statistic $\widehat{Q}$ will be dominated by the contribution from the smallest distance, between z_k and z_l, say. Straightforward calculations give

$$\lim_{d \to 0} S(d) = \begin{cases} \left(\dfrac{N_2(N_2-1)(N-1)}{N_1(N_1-1)(N-2)} \right)^{1/2} & \text{for } k, l \in D \\[2mm] \left(\dfrac{N_1(N_1-1)(N-1)}{N_2(N_2-1)(N-2)} \right)^{1/2} & \text{for } k, l \in D^c \\[2mm] -\left(\dfrac{(N_1-1)(N_2-1)(N-1)}{N_1 N_2(N-2)} \right)^{1/2} & \text{otherwise,} \end{cases} \tag{5.25}$$

where $N = N_1 + N_2$. Under the null hypothesis the above three cases occur with probability $N_1(N_1-1)/(N(N-1))$, $N_2(N_2-1)/(N(N-1))$ and $2N_1 N_2/(N(N-1))$ respectively, and with these probabilities, it can be readily checked that the conditional mean of S is zero and the conditional variance is equal to 1. Subsequently taking the limit $N \to \infty$ with $N_1/N = s$ gives

$$\lim_{\substack{N \to \infty \\ N_1/N = s}} \lim_{d \to 0} S(d) = \begin{cases} (1-s)/s & \text{w.p. } s^2 \\ s/(1-s) & \text{w.p. } (1-s)^2 \\ -1 & \text{w.p. } 2s(1-s). \end{cases} \tag{5.26}$$

For the case $N_1 = N_2$ ($s = 1/2$), one obtains the values 1 and -1 for $\lim_{d \to 0} S(d)$, both with probability $1/2$.

Next, we consider the limit $d \to \infty$. An expression for S in this limit can be found using the relation

$$
\begin{aligned}
h(z_i, z_j) &= e^{-\|z_i - z_j\|^2/(2d^2)} \\
&= 1 - \|z_i - z_j\|^2/(2d^2) + O\left((\tfrac{1}{d})^4\right)
\end{aligned}
\tag{5.27}
$$

which shows that, up to a scale factor which does not enter the ratio S, the statistic Q in this limit is equivalent to the statistic

$$
\begin{aligned}
\widehat{Q}' &= -\frac{1}{\binom{N_1}{2}} \sum\sum_{1 \le i < j \le N_1} \varphi(\boldsymbol{X}_i, \boldsymbol{X}_j) - \frac{1}{\binom{N_2}{2}} \sum\sum_{1 \le i < j \le N_2} \varphi(\boldsymbol{Y}_i, \boldsymbol{Y}_j) \\
&\quad + \frac{2}{N_1 N_2} \sum_{i=1}^{N_1} \sum_{j=1}^{N_2} \varphi(\boldsymbol{X}_i, \boldsymbol{Y}_j)
\end{aligned}
\tag{5.28}
$$

with

$$
\varphi(\boldsymbol{a}, \boldsymbol{b}) = \|\boldsymbol{a} - \boldsymbol{b}\|^2. \tag{5.29}
$$

The conditional mean of this statistic is again zero, and an expression for its conditional variance analogous to Equation (5.9) can be derived. Like for the reversibility test, it is not necessarily true that the test has power against all alternatives in the limit $d \to \infty$, as this property was derived for finite bandwidth values only.

5. Discussion

In this chapter we have proposed a test for the null hypothesis that two probability measures are identical, and subsequently adapted the method so that it can be applied to the empirical reconstruction measures of two time series. The test is based on the statistic $\widehat{Q}$ which is an unbiased estimator of the square of a distance between the two probability measures. The approach is similar to that followed in Chapter 4, where a distance based test statistic was used to detect differences in forward and time-reversed reconstruction measures. The standard deviation of $\widehat{Q}$ under the null hypothesis, and under the assumption of statistical independence of the vectors, is calculated conditionally on the observed set of vectors. For applications to reconstruction measures, a block method is proposed which turns out to be fairly robust with respect to dependence.

The test is related to permutation tests in that the variance under the null hypothesis is calculated by taking into consideration all permutations of the pool of sampled vectors from both sets are equally likely under the null hypothesis. Like permutation tests, the test is nonparametric, because it does not rely on estimates of any model parameters, but unlike common rank tests for univariate data the test is not distribution-free because it is not invariant under invertible transformations of the sample space. A distribution-free multivariate two-sample test, which, however, appears to be difficult to apply in practice, was proposed by Bickel (1969). Several other statistics have been proposed for the multivariate two-sample problem, but critical values of these are often difficult to obtain (Romano; 1988).

Kantz (1994) considered Equation (5.4) with the kernel $\Theta(d - \|\boldsymbol{r} - \boldsymbol{s}\|)$, where $\Theta(\cdot)$ is the Heaviside step function. For this kernel, one can find the conditional variance in a way completely analogous to the method used for our estimator. However, the kernel does not define an estimator with an expected value which may be interpreted as the square of a distance between two probability measures. This is the case only in the limit $d \to 0$ assuming that the reconstruction measures are smooth. Formally, this implies that a test based on this kernel is not consistent against all alternatives. From a practical point of view, however, this kernel has the advantage of being computationally less demanding.

A related test statistic discussed by Anderson *et al.* (1994) is

$$T = \int (\widehat{f_1}(\boldsymbol{x}) - \widehat{f_2}(\boldsymbol{x}))^2 \, \mathrm{d}\boldsymbol{x}, \tag{5.30}$$

where $\widehat{f_1}$ and $\widehat{f_2}$ are kernel estimators of the probability density functions (which are assumed to exist) associated with μ_1 and μ_2 respectively. This test statistic is biased for finite sample sizes, but the problem of estimating the bias is overcome by the use of a bootstrap.

When the test results indicate that two time series are different, one may proceed by trying to identify the sense in which they differ. If the time series differ only in their mean, but are otherwise similar, one expects that the null hypothesis is no longer rejected when the means of both time series are made identical. If there is still a difference between the time series after correcting for their means, one could apply the test again after scaling the time series so that not only their means but also their variances are equal. Such a procedure is also advisable if one is interested in testing the null hypothesis that the time series are identical up to certain differences such as differences in the mean and variance, which are 'trivial' in the sense that one would say that two processes differing only in mean and variance represent equivalent dynamics.

The test can be used to compare empirical time series with surrogate time series, such as phase randomized time series (Theiler *et al.*, 1992a, 1992b). Strictly speaking, a phase randomized surrogate time series and the original time series are not independent since by construction they have identical sample power spectra, but this is expected to have only a small effect on the distribution of S. We expect that the comparison of different segments of a given time series with the proposed test is a promising means of testing for stationarity of an observed time series.

CHAPTER 6

ESTIMATING INVARIANTS OF NOISY ATTRACTORS

This chapter describes a method for estimating the correlation dimension and correlation entropy of a time series based on a Gaussian kernel version of the correlation integral. A simple analytic form is derived for Gaussian kernel correlation integrals of time series corrupted with Gaussian measurement noise. Applications to simulated time series indicate that reasonable estimates of the noise level, the correlation dimension and the correlation entropy can be obtained for time series with up to 20% noise, and that the method is fairly robust with respect to the noise distribution. Applications to empirical time series are given, and analytic expressions for the coarse-grained correlation dimensions and entropies derived.

1. Introduction

The analysis of nonlinear time series in the presence of measurement noise, or observational noise, is a problem of great current interest. For noise-free deterministic time series, powerful methods are available for the estimation of dynamical invariants such as the correlation dimension and the correlation entropy. However, as shown in Chapter 3, measurement noise, which is present in any experimental time series and especially in many physiological time series, can put severe limitations on the estimation of dynamical invariants from time series with these methods.

It will be recalled from Chapter 2 that the Grassberger-Procaccia method allows the determination of both the correlation dimension D_2 and the correlation entropy K_2 of an attractor after constructing the m-dimensional delay vectors $\boldsymbol{X}_n = (X_n, X_{n+\tau}, \ldots, X_{n+(m-1)\tau})$ with delay τ from a time series $\{X_n\}_{n=1}^N$. The correlation integral $C_m(r)$ of the reconstruction measure μ_m is given by

$$C_m(r) = \int\int \Theta(r - \|\boldsymbol{x} - \boldsymbol{y}\|) \, \mathrm{d}\mu_m(\boldsymbol{x}) \, \mathrm{d}\mu_m(\boldsymbol{y}). \tag{6.1}$$

Estimates of D_2 and K_2 are obtained by selecting a scaling region (a range of r-values) in which the model

$$C_m(r) \sim r^{D_2} e^{-m\tau K_2} \tag{6.2}$$

is fitted to the sample correlation integrals $\widehat{C}_m(r)$. For small values of r, measurement noise gives rise to an increased slope in the log-log plot of the correlation integral. For low noise levels, a reasonable scaling region can usually still be found, whereas this is impossible if the noise level is high.

Basically two approaches to analyzing time series with measurement noise can be distinguished. The first is to separate the noise and the underlying time series with a noise reduction method (for surveys see Grassberger et al.a , 1993; Kostelich and Schreiber, 1993). The main idea behind these methods is the use of information from near neighbors under the assumption that the clean evolution is given by a continuous map. The second approach is based on characterizing the noisy reconstruction measure directly through a modification of the scaling law for the correlation integral. By calculating the effect of noise on the correlation integral, Schouten et al. (1994a) obtained a model for the correlation integral in the presence of bounded independent, identically distributed (i.i.d.) noise, from which the noise level and the correlation integral can be estimated. Schreiber (1993a) proposed a method for estimating the noise level of a deterministic time series contaminated with unbounded i.i.d. Gaussian measurement noise. The effect of this noise on the correlation integral was also investigated by Smith (1992b), who derived an expression for the correlation integral and used this to estimate D_2 for small noise levels. The analytic difficulties which prevent the estimation of invariants at higher noise levels appear to be related to the contrast between the smooth Gaussian noise distribution on the one hand and the abrupt nature of the kernel function $\Theta(\cdot)$ in the correlation integral on the other hand. Oltmans and Verheijen (1997) also examined the influence of Gaussian noise on the density of inter-point distances in the attractor (the derivative of the correlation integral) and obtained an expression of the correlation integral similar to that of Smith (1992b) under more general assumptions.

In this chapter we will show that, in the context of i.i.d. Gaussian noise, a more natural formalism is obtained by examining a function from the same family as the correlation integral, but which is tailored for Gaussian measurement noise. We start by considering the correlation integral as a member of a generalized class of kernel integrals. Then a Gaussian kernel member is picked from this class and its behavior is derived analytically in the presence of Gaussian measurement noise. We then give some example applications to noisy deterministic time series for which D_2, K_2 and the noise level σ are estimated.

2. A Generalized Correlation Integral

The correlation integral defined by Equation (6.1) can be generalized to

$$T_m(h) = \int\int K(\|\boldsymbol{x} - \boldsymbol{y}\|/h) \, \mathrm{d}\mu_m(\boldsymbol{x}) \, \mathrm{d}\mu_m(\boldsymbol{y}), \tag{6.3}$$

where $K(\cdot)$ is a kernel function and h is a parameter determining the width of the kernel. The usual correlation integral is retained when the kernel function $K(x)$ is taken to be $\Theta(1 - x)$. The argument h will be referred to as the bandwidth parameter. For the correlation integral, the bandwidth parameter can be interpreted as a radius; the correlation integral is the relative number of inter-point vectors $(\boldsymbol{X}_i - \boldsymbol{X}_j)$ within a ball of radius h. In order to avoid confusion with a radius in the generalized case, we use h rather than r in definition (6.3).

Using the Euclidean norm together with the Gaussian kernel function

$$K(x) = e^{-x^2/4}, \tag{6.4}$$

a version of the correlation integral,

$$T_m(h) = \int\int e^{-\|\boldsymbol{x}-\boldsymbol{y}\|^2/(4h^2)} \, \mathrm{d}\mu_m(\boldsymbol{x}) \, \mathrm{d}\mu_m(\boldsymbol{y}), \tag{6.5}$$

is obtained which will be referred to as the Gaussian kernel correlation integral. Ghez *et al.* (1993) and Ghez and Vaienti (1992a) used a Gaussian kernel function for the estimation of dimensions and entropies of noise-free time series. Gaussian kernel functions were used implicitly in statistical tests for reversibility (Diks *et al.*, 1995) and for comparing the reconstruction measures of two time series (Diks *et al.*, 1996), see chapters 4 and 5 respectively.

First the noise-free scaling law for the Gaussian kernel correlation integrals is established. If we take m fixed and consider a deterministic time series with correlation dimension D_2, then the scaling law

$$T_m(h) \sim h^{D_2}, \qquad \text{for } m \text{ fixed, } h \to 0, \tag{6.6}$$

holds. More generally, scaling according to Equation (6.6) is found for correlation integrals based on any other kernel function $K(x)$ which decreases monotonically in x for $x \geq 0$, and for which

$$\lim_{h \to 0} h^{-p} K(x/h) = 0 \tag{6.7}$$

point-wise for $x > 0$ for all $p \geq 0$ (Ghez and Vaienti, 1992a).

The double integral in Equation (6.5) can be replaced by a single integral over the distribution of distances, which leads to the expression of $T_m(h)$ as

$$T_m(h) = \int \eta_m(r) K(r/h) \, \mathrm{d}r, \qquad (6.8)$$

where $\eta_m(r) = \mathrm{d}C_m(r)/\mathrm{d}r$, the derivative of the usual (Θ-kernel) correlation integral, is the density of the inter-point distances r. Using this expression the dependence on m of $T_m(h)$ can be found as follows. Frank $et~al.$ (1993) showed that the correlation integral calculated with the Euclidean norm behaves as

$$C_m(r) \sim e^{-m\tau K_2} (r/\sqrt{m})^{D_2}, \qquad \text{for } r \to 0,\, m \to \infty, \qquad (6.9)$$

which implies $\eta_m(r) \sim e^{-m\tau K_2} m^{-D_2/2}$ for r fixed. Together with Equation (6.8) this gives

$$T_m(h) \sim e^{-m\tau K_2} m^{-D_2/2}, \qquad \text{for } h \text{ fixed}, \, m \to \infty. \qquad (6.10)$$

Upon combining the two scaling rules for $T_m(h)$ given in Equations (6.6) and (6.10) we find

$$T_m(h) \sim e^{-m\tau K_2} m^{-D_2/2} h^{D_2}, \qquad \text{for } h \to 0,\, m \to \infty \qquad (6.11)$$

for Gaussian kernel correlation integrals in the noise-free case with the Euclidean norm.

Following Frank $et~al.$, we could remove the factor $m^{-D_2/2}$ in Equation (6.11) by defining an m-dependent bandwidth. There is, however, a practical reason for not using this freedom and proceeding with Equation (6.11). Due to the boundedness of the attractor there usually is an upper bandwidth up to which the behavior described by Equation (6.11) is observed, which is approximately independent of m; with dimension scaled bandwidths we would have to use smaller upper bandwidths for increasing m.

3. The Effect of Noise on the Gaussian Kernel Correlation Integral

Note that the scaling law for the Gaussian kernel correlation integral given in Equation (6.11) only holds for noise-free time series. This section discusses the modifications to this scaling law when Gaussian measurement noise is present.

The Gaussian kernel correlation integral can be expressed as

$$T_m(h) = (2h\sqrt{\pi})^m \int (\rho_m^h(x))^2 \, \mathrm{d}x, \qquad (6.12)$$

where ρ_m^h is the density obtained by convolution of μ_m with a spherically symmetrical Gaussian probability density function parameterized by h,

$$\rho_m^h(\boldsymbol{x}) = (h\sqrt{2\pi})^{-m} \int e^{-\|\boldsymbol{x}-\boldsymbol{y}\|^2/(2h^2)} \, \mathrm{d}\mu_m(\boldsymbol{y}). \tag{6.13}$$

This is best demonstrated by deriving Equation (6.5) from Equations (6.12) and (6.13). After substituting Equation (6.13) into Equation (6.12) one obtains

$$T_m(h) = (h\sqrt{\pi})^{-m} \iiint e^{-(\|\boldsymbol{x}-\boldsymbol{y}\|^2+\|\boldsymbol{x}-\boldsymbol{z}\|^2)/(2h^2)} \, \mathrm{d}\mu_m(\boldsymbol{y}) \, \mathrm{d}\mu_m(\boldsymbol{z}) \, \mathrm{d}\boldsymbol{x}. \tag{6.14}$$

After substitution of

$$\|\boldsymbol{x} - \boldsymbol{y}\|^2 + \|\boldsymbol{x} - \boldsymbol{z}\|^2 = 2\sum_{i=1}^{m}(x_i - \frac{1}{2}(y_i + z_i))^2 + \frac{1}{2}\sum_{i=1}^{m}(y_i - z_i)^2, \tag{6.15}$$

in the exponent of the integrand, the integral over $\boldsymbol{x}$ is easily calculated upon introduction of the variable $\boldsymbol{x}' = \boldsymbol{x} - \frac{1}{2}(\boldsymbol{y} + \boldsymbol{z})$ and results in Equation (6.5).

The density $\rho_m^h(\boldsymbol{x})$ defined in Equation (6.13) allows an interpretation as a convolution of the measure μ_m of reconstruction vectors with a Gaussian probability density function with covariance matrix $h^2 I$. This property may be exploited in the presence of Gaussian measurement noise which itself acts on the delay vector distribution as a convolution with a Gaussian.

To see how i.i.d. Gaussian measurement noise affects $T_m(h)$ it is important to make a clear distinction between the reconstruction measure μ_m associated with the noisy reconstruction vectors and the measure $\bar{\mu}_m$ of the underlying noise-free reconstruction vectors. The density $\rho_m(\boldsymbol{x})$ of μ_m can be interpreted as the convolution of $\bar{\mu}_m$ with a probability density function with covariance matrix $\sigma^2 I$ (Casdagli *et al.*, 1991). The two consecutive convolutions can be summarized by the single convolution

$$\rho_m^h(\boldsymbol{x}) = (s\sqrt{2\pi})^{-m} \int e^{-\|\boldsymbol{x}-\boldsymbol{y}\|^2/(2s^2)} \, \mathrm{d}\bar{\mu}_m(\boldsymbol{y}), \tag{6.16}$$

where $s^2 = h^2 + \sigma^2$. Upon substitution of Equation (6.16) into Equation (6.12) and expressing the double integral in the in the form of Equation (6.5), one finds

$$T_m(h) = \sqrt{\frac{h^2}{h^2 + \sigma^2}}^{-m} \iint e^{-\|\boldsymbol{x}-\boldsymbol{y}\|^2/(4h^2+4\sigma^2)} \, \mathrm{d}\bar{\mu}_m(\boldsymbol{x}) \, \mathrm{d}\bar{\mu}_m(\boldsymbol{y}), \tag{6.17}$$

which describes $T_m(h)$ in terms of the underlying clean reconstruction measure $\bar{\mu}_m$ in the presence of i.i.d. Gaussian noise with standard deviation σ.

The behavior of the double integral in Equation (6.17) is found from the definition of $T_m(h)$ given in Equation (6.5) together with the noise-free scaling law (6.11), leading to

$$T_m(h) \simeq \phi \sqrt{\frac{h^2}{h^2 + \sigma^2}}^{-m} e^{-m\tau K_2} m^{-D_2/2} \sqrt{h^2 + \sigma^2}^{D_2}$$

$$(6.18)$$

$$\text{for } \sqrt{h^2 + \sigma^2} \to 0, \; m \to \infty,$$

where ϕ is a normalization constant. It can be readily verified that taking the limit $\sigma \to 0$ gives the noise-free scaling relation in Equation (6.11). Note that the Gaussian kernel correlation integral for small values of h and m fixed behaves as $T_m(h) \sim h^m$, which is a manifestation of the m-dimensionality of the set of noisy delay vectors. In practice the standard deviation σ of the noise level is fixed at a nonzero value, so that we can not let $\sqrt{h^2 + \sigma^2}$ tend to zero. Nevertheless, we expect Equation (6.18) to hold good in a range of small values of h provided that the noise level σ is not too large.

4. Applications to Simulated Time Series

In all applications the time series are rescaled so that their sample standard deviation equals 1. The quoted noise levels are noise levels after rescaling which allows for a convenient comparison of the bandwidth parameter h and the noise levels σ for the different time series considered. The case $\sigma = 0$ corresponds to a clean noise-free time series while a noise level of $\sigma = 1$ implies a time series consisting of Gaussian i.i.d. random variables. Denoting the variance of the original time series by r^2, and the variance of the noise by s^2, we have

$$\sigma^2 = \frac{s^2}{r^2 + s^2},$$

$$(6.19)$$

so that time series with a given normalized noise level σ can be obtained by adding noise with a variance

$$s^2 = r^2 \frac{\sigma^2}{1 - \sigma^2}$$

$$(6.20)$$

to clean time series with variance r.

The Gaussian kernel correlation integrals $T_m(h)$ can be estimated consistently by their U-statistic estimator, obtained by averaging the Gaussian kernel function (the integrand in Equation 6.5) over pairs of delay vectors. The

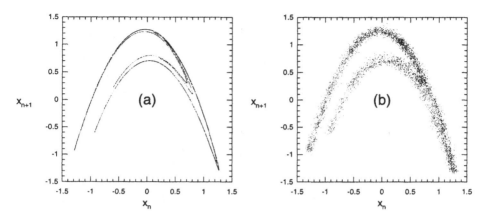

Figure 6.1: *Phase portraits for a noise-free ($\sigma = 0$) Hénon time series (a) and a noisy Hénon time series with $\sigma = 0.05$ (b).*

estimator $\widehat{T}_m(h)$ becomes

$$\widehat{T}_m(h) = \frac{1}{N_p} \sum_{i \in R} \sum_{j \neq i} w_{ij}(h), \qquad (6.21)$$

where R is a set of reference indices chosen randomly without replacement, N_p is the total number of (i, j) pairs summed over and

$$w_{ij}(h) = e^{-\|\boldsymbol{X}_i - \boldsymbol{X}_j\|^2/(4h^2)} \qquad (6.22)$$

is the relative contribution of the pair of points $(\boldsymbol{X}_i, \boldsymbol{X}_j)$. To reduce the effect of dependence among delay vectors one can easily incorporate a Theiler correction by not taking into account pairs of points that are close in time (cf. Theiler, 1986).

We applied our method to various Hénon time series with noise levels σ ranging from 0 to 0.20. To illustrate the difference in appearance of a clean and a noisy attractor, Figure 6.1 shows phase portraits of the clean attractor ($\sigma = 0$) and of the attractor with a normalized noise level of $\sigma = 0.05$. The self-similar structure of the attractor is no longer visible at this noise level. Throughout, the Gaussian kernel correlation integrals are calculated for bandwidth values that are equidistant on a logarithmic scale, with 2 values per binade. This choice has the advantage that we only need to perform the numerically time-consuming evaluation of the exponential function in $w_{ij}(h)$ for the largest value of the bandwidth parameter. The values of

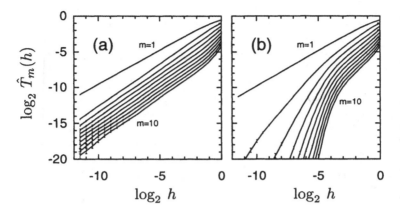

Figure 6.2: *Estimated Gaussian kernel correlation integrals $\widehat{T}_m(h)$ as a function of the bandwidth h on a log-log scale for a Hénon time series without noise (a) and with a (normalized) noise level of 0.05 (b). The different lines correspond to the cases $m = 1$ (upper line) up to $m = 10$ (lower line). The bars denote the estimated standard error.*

$w_{ij}(h)$ at the smaller bandwidths can then be found efficiently using the relation $w_{ij}(h/\sqrt{2}) = w_{ij}^2(h)$.

Figure 6.2(a) shows the estimated Gaussian kernel correlation integrals $\widehat{T}_m(h)$ obtained using 10^3 random reference points as a function of the bandwidth h on a log-log scale for a noise-free time series generated with the Hénon model, of length $N = 4,000$. As the noise-free scaling relation in Equation (6.11) suggest, the curves in Figure 6.2(a) are linear and parallel for small values of h and large values of m. Figure 6.2(b) shows a log-log plot of $\widehat{T}_m(h)$ versus h for the same Hénon time series with i.i.d. Gaussian noise with a standard deviation of $\sigma = 0.05$. It can be observed in Figure 6.2(b) that the noise gives rise to an increased slope for small bandwidth values h.

For various values of the noise level, a Marquardt nonlinear fit procedure (Press *et al.*, 1992) for the parameters ϕ, K_2, D_2 and σ was performed in the range where $h \leq 0.25$, and $\widehat{T}_m(h) > 2/N_{\mathrm{p}}$. For each m, the model given in Equation (6.18) was fitted simultaneously to the sample Gaussian kernel correlation integrals $\widehat{T}_m(h)$ and $\widehat{T}_{m+1}(h)$. The lower bound on $\widehat{T}_m(h)$ in the fit is imposed to avoid situations in which $\widehat{T}_m(h)$ is dominated by a single inter-point distance. As each distance has a contribution of at most $1/N_{\mathrm{p}}$, as a heuristic rule, we put this lower bound on $2/N_{\mathrm{p}}$. Note that a similar bound is usually put implicitly on fits of the traditional correlation integral. The

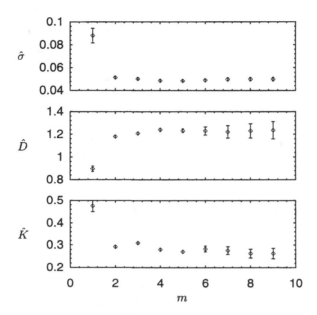

Figure 6.3: *Estimated values $\hat{\sigma}$, $\hat{D}_2$ and $\hat{K}_2$ as a function of m for a Hénon time series with a noise level σ of 0.05. The bars denote the estimated 95% confidence interval (2 standard errors).*

correlation integral can not be estimated well below the smallest inter-point distance, $r_{\min}$, available in the cloud of delay vectors. The condition $r > r_{\min}$ usually put on the range of radii for fits of the correlation integral implies $C_m(r) > 1/N_p$.

Assuming independence of the distances $\|\boldsymbol{X}_i - \boldsymbol{X}_j\|$, the variance of $\hat{T}_m(h)$ can be estimated consistently as

$$Var(\hat{T}_m(h)) = \frac{1}{N_p}(\overline{w_{ij}^2(h)} - \overline{w_{ij}(h)}^2), \tag{6.23}$$

where the bars denote averaging over the pairs (i,j). These estimated standard deviations were used as weights in the fit procedure. The estimates of the model parameters at a noise level of $\sigma = 0.05$ as a function of m are shown in Figure 6.3. Reasonable estimates of σ, D_2 and K_2 are obtained at moderately small values of m ($m = 3$ or $m = 4$).

σ	$\hat{\sigma}$	$\hat{D}$	$\hat{K}$
0.00	0.00004 ± 0.00009	1.196 ± 0.002	0.296 ± 0.003
0.01	0.0102 ± 0.00006	1.205 ± 0.002	0.293 ± 0.002
0.02	0.0195 ± 0.0002	1.200 ± 0.004	0.294 ± 0.003
0.05	0.0487 ± 0.0003	1.240 ± 0.007	0.280 ± 0.003
0.10	0.0996 ± 0.0008	1.26 ± 0.02	0.309 ± 0.003
0.20	0.206 ± 0.003	1.16 ± 0.05	0.278 ± 0.003

Table 6.1: *Estimates of the noise level $\hat{\sigma}$, correlation dimension $\hat{D}_2$ and correlation entropy $\hat{K}_2$ for the Hénon attractor with different levels of Gaussian observational noise σ. The values are the estimates at $m = 4$.*

The results for $m = 4$ and different noise levels ranging up to $\sigma = 0.20$ are summarized in Table 6.1. The estimates of σ are close to their true values and for noise levels up to 0.10, the values of $\hat{D}_2$ and $\hat{K}_2$ are close to the values $D_2 \approx 1.22$ and $K_2 \approx 0.29$ nats/t.u. found in the literature (see (Grassberger, 1983), and (Frank *et al.*, 1993) respectively). The estimated values of the standard error, however, seem to be on the small side. This is possibly due to covariance between the estimates of the Gaussian kernel correlation integrals for different values of h and m, which was not taken into account in the fit procedure.

We applied the method to a noise-free time series ($N = 10^4$, sampling time 0.5 s.) generated with the Rössler model (Grassberger *et al.*, 1991). The estimated parameters at $m = 9$ for $\tau = 3$ were $\hat{\sigma} = 0.0007 \pm 0.0002$, $\hat{D}_2 = 1.97 \pm 0.01$ and $\hat{K}_2 = 0.066 \pm 0.009$ nats/t.u. An application to a noisy Rössler time series ($N = 10^4$, sample time 0.5, $\sigma = 0.05$) gave the estimates $\hat{\sigma} = 0.0507 \pm 0.0003$, $\hat{D}_2 = 1.94 \pm 0.03$ and $\hat{K}_2 = 0.06 \pm 0.01$ nats/t.u. The information dimension, which was estimated by Grassberger *et al.* (1991) as 2.00 ± 0.01, is an upper limit for the value of D_2. Using the method of Schouten *et al.* (1994b), we have found a correlation entropy of 0.07 nats/t.u. for the clean Rössler time series. These examples show that our method can also be applied to time series obtained from continuous time dynamical systems.

The sensitivity of our method with respect to the type of measurement noise, was examined by applying it to a Hénon time series with independent uniformly distributed noise with a standard deviation σ of 0.05. The estimated parameters are $\hat{\sigma} = 0.0454 \pm 0.0003$, $\hat{D}_2 = 1.220 \pm 0.007$ and $\hat{K}_2 = 0.294 \pm 0.003$ nats/t.u. Although the estimated noise level is about 10% too small, the estimates of D_2 and K_2 are still very reasonable, which suggests that the method can be used for various noise distributions.

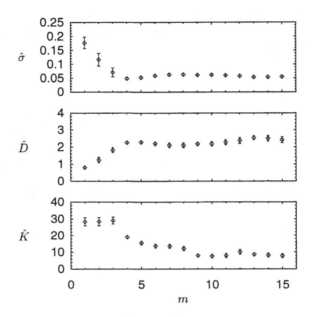

Figure 6.4: *Estimated values $\hat{\sigma}$, $\hat{D}_2$ and $\hat{K}_2$ (in nats/s.) for the atrial fibrillation time series. The bars denote the estimated 95% confidence interval (2 standard errors).*

5. Applications

The following examples describe the results obtained in applications to two physiological and one physical time series.

5.1. Example: Atrial Fibrillation

As a first example we reconsider the atrial fibrillation time series which was analyzed in Chapter 2 (section 3.2) with the Grassberger-Procaccia method. There, the estimated correlation dimension did not saturate but kept increasing slowly with the embedding dimension. For the Gaussian kernel correlation integral method again a delay of 21 sample times was used in the reconstruction. A Theiler correction of twice the delay time was used, and the model was fitted in the bandwidth range [0.01, 0.25]. The estimated noise level and dynamical invariants are shown in Figure 6.4. The correlation dimension of the time series is about 2.5, and the correlation entropy is close to 10 nats/s. These results suggest that some types of atrial fibrillation can be characterized by low-dimensional dynamics.

5.2. Example: Respiration

We applied the method to a recording of the air flow of an adult anesthetized cat. This signal was sampled at 10Hz using a 12-bit ADC. Before analysis we sampled down the time series to a sampling time of 0.5 s. A segment of 100s of the time series is shown in Figure 6.5. Occasionally the signal contains a large spike which is the result of a sigh. A delay of 0.5 s. was used in the reconstruction. The Theiler correction was set to twice the delay time and the model was fitted in the bandwidth range [0.001, 0.25].

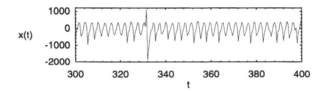

Figure 6.5: *Segment of 100 seconds of cat respiration signal.*

The estimated parameters are shown in Figure 6.4 for increasing values of the embedding dimension. The estimated noise level and correlation entropy saturate at about 0.08 and 0.3 nats/s. respectively, but the estimated correlation dimension keeps increasing with the embedding dimension. The latter is an indication that a description in terms of low-dimensional chaos plus Gaussian observational noise is not appropriate for the time series.

5.3. Example: Fluidized Bed Pressure

We next consider a fluidized bed pressure time series of length $N = 4960$ and sampling time $\Delta t = 0.002$s., kindly provided by the Chemical Reaction Engineering group of Delft University of Technology. Such pressure time series can be expected to have a complex nature as they are measured locally in a system where the spatial aspect is important. The typical length scale of velocity fluctuations in the system is small compared to the system size, but information from anywhere in the system can reach the detector from the entire system through pressure waves in the system. The pressure time series,

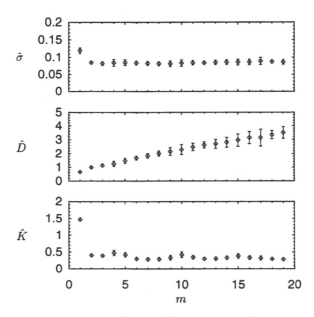

Figure 6.6: *Estimated values $\hat{\sigma}$, $\hat{D}_2$ and $\hat{K}_2$ (in nats/s.) for the respiration signal. The bars denote the estimated 95% confidence interval (2 standard errors).*

a section of which is shown in Figure 6.7, therefore can not even be considered a local time series.

The estimated parameters for the fluidized bed series are shown in Figure 6.8 as a function of m, for a delay $\tau = 15$ and a Theiler correction $T = 2\tau$. The values of $\hat{\sigma}$ and $\hat{K}_2$ appear to converge to the values 0.18 and 11 nats/s. respectively. The estimated value $\hat{D}_2$ of the correlation dimension, however, does not saturate with increasing values of the embedding dimension m. This

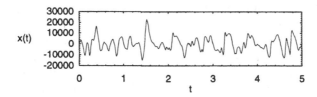

Figure 6.7: *Segment of five seconds of the fluidized bed pressure time series.*

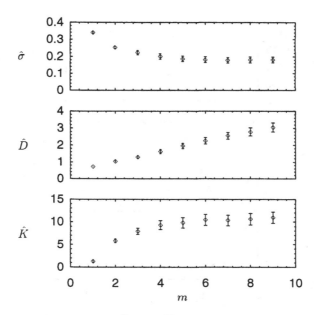

Figure 6.8: *Estimated values $\widehat{\sigma}$, $\widehat{D}_2$ and $\widehat{K}_2$ (in nats/s.) as a function of m for the fluidized bed time series. The bars denote the estimated 95% confidence interval (2 standard errors).*

implies that the model for the Gaussian kernel correlation integral, which is based on the assumption of a low-dimensional deterministic time series plus observational noise, does not describe the data well. We must conclude that the fluidized bed series is not a low-dimensional time series with a relatively small amount of observational noise. We conjecture that dynamical noise plays a large role in the system, and that the nonlocal nature of the pressure time series, together with the high-dimensionality of the system as a whole, also gives rise to a high dimensional pressure time series.

Although the model does not fit the data very well, as the estimates of D_2 for increasing embedding dimensions show, possibly the values to which $\widehat{\sigma}$ and $\widehat{K}_2$ converge as a function of m can still be interpreted as being meaningful characteristics of the time series. To examine this possibility, we repeated the estimation of the model parameters for various different values of the cut-off bandwidth (not shown), and found that the values to which $\widehat{\sigma}$ and $\widehat{K}_2$ converge are dependent on this bandwidth. It thus appears that the method can not be used to characterize the fluidized bed time series, except from making clear that an explanation of the series in terms of low-dimensional chaos plus noise is inappropriate.

6. Coarse-grained Quantities

The scaling law for the usual correlation integral

$$C_m(r) \simeq \phi r^{-D_2} e^{-m\tau K_2} \tag{6.24}$$

for small r and large m suggests that the quantity

$$D_{2,m}(r) = \frac{d \ln C_m(r)}{d \ln r} = \frac{r}{C_m(r)} \frac{d C_m(r)}{dr} \tag{6.25}$$

for small r approaches the correlation dimension D_2, provided that m is sufficiently large. The quantity $D_{2,m}(r)$ at m and fixed nonzero r is known as a resolution dependent effective correlation dimension, or coarse-grained correlation dimension. Similarly,

$$K_{2,m}(r) = \frac{1}{\tau} \left(\ln C_m(r) - \ln C_{m+1}(r) \right) \tag{6.26}$$

defines a coarse-grained version of the correlation entropy. The coarse-grained correlation dimension and entropy are useful for studying correlation integrals which only show approximate scaling, such as those of time series with noise. Deviations from the scaling law should become apparent through the r-dependence of these coarse-grained quantities.

In analogy with the coarse-grained versions of the correlation dimension and correlation entropy for the usual correlation integral, we can define the coarse-grained correlation dimension and entropy of the Gaussian kernel correlation integral. The coarse-grained dimension $D_{2,m}^g(h)$ at embedding dimension m is defined as the derivative of the logarithm of the correlation integral, $\ln T_m(h)$ with respect to the logarithm of the bandwidth, $\ln h$,

$$D_{2,m}^g(h) = \frac{d \ln T_m(h)}{d \ln h}, \tag{6.27}$$

and we define the coarse-grained entropy for the Gaussian kernel correlation integral as

$$K_{2,m}^g(h, \tau) = \frac{1}{\tau} (\ln T_m(h) - \ln T_{m+1}(h)). \tag{6.28}$$

The model of the Gaussian kernel correlation integral of noisy time series, given in Equation (6.18), gives

$$\ln T_m(h) = \ln \phi + \frac{m}{2} \ln \left(\frac{h^2}{h^2 + \sigma^2} \right) - m\tau K_2 - \frac{D_2}{2} \ln m + \frac{D_2}{2} \ln(h^2 + \sigma^2), \quad (6.29)$$

from which we find

$$D_{2,m}^{\text{g}}(h) = \frac{D_2 h^2 + m\sigma^2}{h^2 + \sigma^2}. \quad (6.30)$$

As pointed out by Schreiber (1997) this relation clearly shows a crossover from $D_{2,m}(h) \approx m$ for $h \ll \sigma$ to $D_{2,m} \approx D_2$ for $h \gg \sigma$. Thus on small length scales the noise is dominant and the coarse-grained dimension roughly equals the embedding dimension m, whereas on large length scales the deterministic nature of the time series dominates and gives rise to a coarse-grained dimension equal to D_2.

The coarse-grained entropy of the Gaussian kernel correlation integral reads

$$K_{2,m}^{\text{g}}(h, \tau) = K_2 - \frac{1}{2\tau} \ln \left(\frac{h^2}{h^2 + \sigma^2} \right) + \frac{D_2}{2\tau} \ln \left(\frac{m+1}{m} \right). \quad (6.31)$$

For large values of m the last term vanishes, and the coarse-grained entropy becomes independent of m. In the limit $m \to \infty$ the coarse-grained entropy differs from the true entropy by an amount which depends on the ratio of h and σ. For m large and $h \ll \sigma$ one obtains

$$K_{2,m}^{\text{g}}(h, \tau) \simeq K_2 - \frac{1}{\tau} \ln(h/\sigma). \quad (6.32)$$

Schreiber found that the quantities

$$d_{m,n}(r) = (D_{2,m}(r) - D_{2,n}(r))/(m - n) \quad (6.33)$$

for the usual correlation integral determined with the supremum norm for large m and n have an identical functional dependence on r, the scale of which is determined by the noise level σ. He obtained

$$d_{m,n}(r) = g \left(\frac{r}{2\sigma} \right), \quad (6.34)$$

with

$$g(z) = \frac{2}{\sqrt{\pi}} \frac{z e^{-z^2}}{\text{erf}(z)}, \quad (6.35)$$

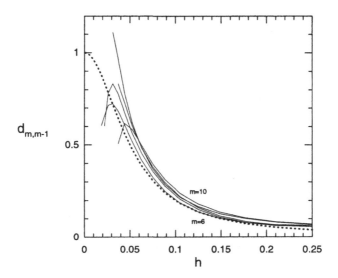

Figure 6.9: *Estimated values of $d^g_{m,m-1}(h)$ vs. h for a Hénon time series with observational noise with $\sigma = 0.05$ for $m = 6, \ldots, 10$ (solid lines) and the function according to the model given in Equation (6.36) for $\sigma = 0.05$ (dashed line).*

which is 1 for $z = 0$ and decreases monotonically to zero for large z. Therefore, $d_{m,n}(r)$ falls off from 1 to zero on a scale that is proportional to the noise level σ.

It can be easily verified that the analogous expression for the Gaussian kernel correlation integral is

$$d^g_{m,n}(h) = f\left(\frac{h}{\sigma}\right), \tag{6.36}$$

with

$$f(z) = \frac{1}{1 + z^2}, \tag{6.37}$$

which is also a sigmoidal function falling off monotonically from 1 to 0 for large z, whence $d^g_{m,n}(h)$ decays from 1 to zero on a scale proportional to the noise level σ.

Figures 6.9 and 6.10 show the estimated values of $d^g_{m,m-1}(h)$ for $m = 6, \ldots, 10$ for a Hénon time series and a Rössler time series respectively, both with noise level $\sigma = 0.05$. The estimated values of $d^g_{m,m-1}(h)$ can be seen to be close to the theoretical curves for $\sigma = 0.05$.

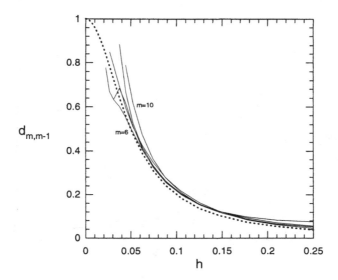

Figure 6.10: *Estimates of $d^g_{m,m-1}(h)$ vs. h for a Rössler time series with observational noise with $\sigma = 0.05$ for $m = 6, \ldots, 10$ (solid lines) and the function according to the model given in Equation (6.36) for $\sigma = 0.05$ (dashed line). The curves of $d^g_{m,m-1}(h)$ are close to the model curve.*

Schreiber estimated the noise level σ of time series by fitting the right hand side of Equation (6.34), which involves only the parameter σ, to estimates of $d_{m,n}(r)$. This enables the estimation of σ independently of D_2 and K_2. Similarly, the right hand side of Equation (6.36) could be fitted to the observed values of $d^g_{m,n}(h)$, giving an estimate of σ independently of D_2 and K_2.

The functional dependence of the coarse-grained quantities on the bandwidth h and the embedding dimension m can also be used to assess the consistency with the assumptions of a chaotic time series with Gaussian observational noise. Figure 6.11 shows the estimates of $d^g_{m,m-1}(h)$ for the fluidized bed time series for $m = 6, \ldots, 10$. The resulting curves as a function of h are not identical (or near identical) as one would expect for a deterministic time series with Gaussian observational noise. We can thus conclude from the estimates of $d^g_{m,m-1}(h)$ that the fluidized bed time series time is not consistent with a low-dimensional time series with Gaussian observational noise.

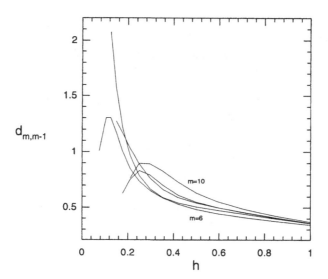

Figure 6.11: *Estimates of $d^g_{m,m-1}(h)$ vs. h for the fluidized bed time series for $m = 6, \ldots, 10$. The curves for different embedding dimensions do not appear to be identical (or nearly identical) as one would expect for a low-dimensional chaotic time series with Gaussian observational noise.*

7. Discussion

In conclusion, we have introduced Gaussian kernel correlation integrals $T_m(h)$, tailored for the characterization of reconstruction measures in the presence of Gaussian measurement noise. The behavior of the Gaussian kernel correlation integrals is derived analytically in terms of the noise level σ and the invariants D_2 and K_2, which enables the estimation of the noise level and the invariants simultaneously using a Marquardt type of nonlinear fit procedure. The estimates of D_2 and K_2 obtained with maps and continuous time dynamical systems agree well with the noise-free values. The standard errors of the parameters appear to be somewhat underestimated, possibly due to the neglected covariance among the estimates of the Gaussian kernel correlation integrals for different values of h and m.

The numerical results suggest that the method can handle noise levels of up to 20% in standard deviation, and also gives reasonable estimates for the dimension and entropy of time series with uniformly distributed noise although the noise level is slightly underestimated in that case. For non-Gaussian noise distributions we expect an improvement with the use of linear combinations of several coordinates like in reconstructions based on singular value decompositions. We furthermore expect that the requirement that the noise be i.i.d. can be relaxed when a delay time τ of the order of the autocorrelation time of the noise is used.

In practice an appropriate choice of the upper cut-off bandwidth has to be made. This is not as straightforward as in the noise-free case where the selection of a scaling region by visual inspection is relatively straightforward. The method is based on the assumptions of low-dimensional determinism and Gaussian measurement noise, but in practice it is usually not known whether this is justified. To prevent spurious estimates of the dynamical invariants and the noise level, the quality of the fit below the upper bandwidth chosen should be investigated and the stability of the estimated parameters upon changing the embedding parameters τ and m assessed.

CHAPTER 7

THE CORRELATION INTEGRAL OF NOISY ATTRACTORS

An analytic expression for the correlation integral is obtained for deterministic time series with Gaussian observational noise. The expression is found by establishing the relation between the correlation integral and the Gaussian correlation integral, for which an analytic expression was already derived in the previous chapter. Our results are consistent with those of Smith (1992a) and Oltmans and Verheijen (1997) who derived expressions for the correlation integral in the presence of noise using different approaches.

1. Introduction

The effect of observational noise on the correlation integral has been studied by several authors, including Ott and Hanson (1981), Ott *et al.* (1985), Smith (1992a), Schreiber (1993a) and Oltmans and Verheijen (1997). In this chapter we will derive an expression for the correlation integral in the presence of Gaussian observational noise and compare it with the results available in the literature.

It will be recalled that the correlation integral $C_m(r)$ of the reconstruction measure μ_m associated with the m-dimensional delay vectors is defined as the probability that the distance between a pair of vectors independently chosen according to μ_m is smaller than r. The correlation integral of deterministic time series exhibits a scaling law

$$C_m(r) \sim e^{-m\tau K_2} r^{D_2}, \qquad (7.1)$$

for r small and m large, with the correlation dimension D_2 and the correlation entropy K_2 as parameters. The values of D_2 and K_2 are usually estimated from the sample correlation integral by fitting this scaling law in a specified region of r-values, called a scaling region. In the presence of Gaussian observational noise the scaling law no longer holds, and estimates of D_2 and K_2 based on it are unreliable. One can therefore ask whether a modified law for the behavior of the correlation integral can be found in the presence of observational noise. Ideally, one could then use this adapted model to estimate D_2 and K_2 and the noise level σ.

The following closed form expression for the correlation integral $C_m(r)$ in the presence of Gaussian observational noise was derived by Smith (1992a)

$$C_m(r) = ar^{D_2} \left(\frac{r^2}{4\sigma^2}\right)^{(m-D_2)/2} \frac{\Gamma(D_2/2+1)}{\Gamma(m/2+1)} M\left(\frac{m-D_2}{2}, \frac{m}{2}+1, -\frac{r^2}{4\sigma^2}\right),$$
(7.2)

where σ is the noise level and a is a factor which is independent of the embedding dimension m and the radius r. M is Kummer's confluent hypergeometric function which has the integral representation

$$M(a, b, z) = \frac{\Gamma(b)}{\Gamma(b-a)\Gamma(a)} \int_0^1 e^{zt} t^{a-1}(1-t)^{b-a-1} \, dt,$$
(7.3)

for Re $b >$ Re $a > 0$ (Abramowitz and Stegun, 1965). Equation (7.2) was derived by Smith under the assumption that the clean attractor underlying the noisy set of delay vectors has an integer-valued dimension D_2. This excludes cases where the clean attractor has a fractal correlation dimension. However, the result appears to be valid also, at least to a good approximation, for attractors with fractal structure. Smith showed that a maximum likelihood procedure for estimation of the correlation dimension D_2, based on Equation (7.2), works well for noisy Hénon time series, characterized by a fractal correlation dimension of about 1.2. This can be explained by the results of Oltmans and Verheijen (1997) who derived the same expression under general conditions.

The aim of this chapter is to derive an expression for the correlation integral from the closed form expression for the Gaussian kernel correlation integral in the presence of Gaussian observational noise found in the previous chapter. First the closed form expression for the Gaussian kernel correlation integral is linked to the usual correlation integral using Laplace transforms. Then the resulting expression for the correlation integral is compared to Equation (7.2).

2. Density of Distances

This section considers the density $\rho(r)$ of distances between reconstruction vectors for a deterministic time series with Gaussian observational noise. Notice that this density always exists, even if the underlying reconstruction measure of the clean time series has point masses or is fractal, because the density of the noise-corrupted delay vector can be expressed as the convolution of the underlying measure of the clean delay vectors with a (smooth) multivariate Gaussian measure.

The correlation integral (Equation (2.20)) can alternatively be written as

$$C_m(r) = \iint K_\theta \left(\frac{\|x - y\|}{r} \right) d\mu(x)\, d\mu(y), \qquad (7.4)$$

where $K_\theta(\cdot)$ is the kernel function given by

$$K_\theta(s) = \Theta(1 - s). \qquad (7.5)$$

In Chapter 6 we introduced the Gaussian kernel correlation integral $T_m(h)$ at embedding dimension m and bandwidth h, defined as

$$T_m(h) = \iint K \left(\frac{\|x - y\|}{h} \right) d\mu(x)\, d\mu(y), \qquad (7.6)$$

where $K(\cdot)$ is a Gaussian kernel function,

$$K(s) = e^{-s^2/4}. \qquad (7.7)$$

Notice the similarity between the definition of the Gaussian kernel correlation integral and the usual definition of the correlation integral. They both are defined as averages of kernel functions, and they both have a parameter which controls the characteristic length scale, or width, of the kernel. Rather than using a kernel function which jumps abruptly from 1 to 0 at a certain radius r, the Gaussian kernel correlation integral uses a smooth Gaussian kernel which decreases smoothly from 1 to 0, on a scale characterized by the width parameter, or bandwidth, h.

In the previous chapter it was shown that the Gaussian kernel correlation integral is particularly useful for estimating dynamical invariants of deterministic time series contaminated with Gaussian observational noise. A closed form expression (Equation 6.17) for the Gaussian kernel correlation integral was obtained, containing the correlation dimension D_2, the correlation entropy K_2 and the variance of the noise, σ^2, as parameters. We have

$$T_m(h) \simeq \phi \sqrt{\frac{h^2}{h^2 + \sigma^2}}^{\,m} e^{-m\tau K_2} m^{-D_2/2} \sqrt{h^2 + \sigma^2}^{\,D_2}, \qquad (7.8)$$

where τ is the delay time used in the embedding, m is the embedding dimension and ϕ is a normalization constant which may depend on the dynamical system at hand and on the delay τ, but not on h or m.

The Gaussian kernel correlation integral can be expressed in terms of the density of distances $\rho(r)$ through

$$T_m(h) = \int_0^\infty \rho(r)e^{-r^2/(4h^2)}\,dr. \tag{7.9}$$

The density $\rho(r)$ can be found from $T_m(h)$ by inverting Equation (7.9) for the particular case where $T_m(h)$ is of the form described by Equation (7.8). Upon substitution of $r = 2\sqrt{t}$ and $s = 1/h^2$ Equations (7.8) and (7.9) give

$$\phi e^{-m\tau K_2}m^{-D_2/2}\sigma^{D_2-m}s^{-D_2/2}(\sigma^{-2} + s)^{(D_2-m)/2} = \int_0^\infty \nu(t)e^{-st}\,dt \tag{7.10}$$

with

$$\nu(t) = \frac{2\rho(2\sqrt{t})}{\sqrt{t}}, \tag{7.11}$$

which may be interpreted as the density of $t = r^2/4$. Notice that the right hand side of Equation (7.10) is just the Laplace transform of $\nu(t)$. Because a solution of Equation (7.10) in terms of $\nu(t)$ will provide an expression for $\rho(r)$ upon inverting Equation (7.11), we first concentrate on finding the inverse Laplace transform of the left hand side of Equation (7.10).

We write the left hand side of Equation (7.10) as $pf_m(s)g_m(s)$, where

$$p = \phi e^{-m\tau K_2}m^{-D_2/2}\sigma^{D_2-m}, \tag{7.12}$$

is an s-independent factor,

$$f_m(s) = s^{-D_2/2} \qquad \text{and} \qquad g_m(s) = \left(\frac{1}{\sigma^2} + s\right)^{\frac{D_2-m}{2}}. \tag{7.13}$$

The inverse Laplace transforms of $f_m(s)$ and $g_m(s)$ are denoted by their upper case analogs, $F_m(t)$ and $G_m(t)$ respectively. It follows from Equation (29.2.8) of Abramowitz and Stegun (1965) that the inverse Laplace transform of the product $f_m(s)g_m(s)$ is the convolution of the inverse Laplace transforms $F_m(t)$ and $G_m(t)$. The density $\nu(t)$ thus is p times the convolution of F_m and G_m:

$$\nu(t) = p \times (F_m * G_m)(t). \tag{7.14}$$

Using Equation (29.3.7) of Abramowitz and Stegun (1965) one finds:

$$F_m(t) = \frac{t^{D_2/2-1}}{\Gamma(\frac{D_2}{2})} \qquad \text{and} \qquad G_m(t) = e^{-t/(\sigma^2)}\frac{t^{(m-D_2)/2-1}}{\Gamma(\frac{m-D_2}{2})}. \tag{7.15}$$

After expanding the convolution of F_m and G_m explicitly and introducing the variable $q = \tau/t$ the integral can be rewritten as

$$(F_m * G_m)(t) = \frac{t^{m/2-2}}{\Gamma(\frac{D_2}{2})\Gamma(\frac{m-D_2}{2})} \int_0^1 (1-q)^{D_2/2-1} q^{(m-D_2)/2-1} e^{-\tau q/(\sigma^2)} t\, dq,$$

(7.16)

and with Equation (13.2.1) of Abramowitz and Stegun (1965) this can be written as

$$(F_m * G_m)(t) = t^{m/2-1} \frac{1}{\Gamma(\frac{m}{2})} M\left(\frac{m - D_2}{2}, \frac{m}{2}, -\frac{t}{\sigma^2}\right).$$

(7.17)

For $\nu(t)$ we thus find

$$\nu(t) = \phi e^{-m\tau K_2} m^{-D_2/2} \sigma^{D_2-m} t^{m/2-1} \frac{1}{\Gamma(\frac{m}{2})} M\left(\frac{m - D_2}{2}, \frac{m}{2}, -\frac{t}{\sigma^2}\right),$$

(7.18)

which finally provides us with the following expression for the density $\rho(r)$ of distances:

$$
\begin{aligned}
\rho(r) \;=\; & \phi e^{-m\tau K_2} m^{-D_2/2} 2^{1-D_2} r^{D_2-1} \left(\frac{r^2}{4\sigma^2}\right)^{\frac{m-D_2}{2}} \\
& \times \frac{1}{\Gamma(\frac{m}{2})} M\left(\frac{m - D_2}{2}, \frac{m}{2}, -\frac{r^2}{4\sigma^2}\right).
\end{aligned}
$$

(7.19)

In the next section it is shown that this expression is consistent with the expression of the correlation integral given in Equation (7.2) for a particular choice of a.

3. Correlation Integral

The density ρ of distances between pairs of points taken according to the reconstruction measure is the derivative of the correlation integral with respect to r, which follows from the fact that the correlation integral $C_m(r)$ can be expressed as the integral of the density of distances over the range of distances from 0 to r:

$$C_m(r) = \int_0^r \rho(v)\, dv.$$

(7.20)

We will take the derivative with respect to r of the expression for the correlation integral given in Equation (7.2), and then try to choose a consistently with

Equations (7.19) and (7.20). If this can be done, it follows that Equation (7.2) and our approach are consistent.

Equation (7.2) gives

$$
\frac{dC_m(r)}{dr} = ar^{D_2-1}\left(\frac{r^2}{4\sigma^2}\right)^{\frac{m-D_2}{2}}\frac{\Gamma(D_2/2+1)}{\Gamma(m/2+1)}
$$

$$
\times\left\{mM\left(\frac{m-D_2}{2},\frac{m}{2}+1,-\frac{r^2}{4\sigma^2}\right)\right. \tag{7.21}
$$

$$
\left.+r\frac{d}{dr}M\left(\frac{m-D_2}{2},\frac{m}{2}+1,-\frac{r^2}{4\sigma^2}\right)\right\}.
$$

The term within the curly braces, which will be denoted by P, can be simplified by first substituting $z = -r^2/(4\sigma^2)$. We obtain,

$$
P = mM\left(\frac{M-D_2}{2},\frac{m}{2}+1,z\right)+2z\frac{d}{dz}M\left(\frac{M-D_2}{2},\frac{m}{2}+1,z\right). \tag{7.22}
$$

Using Equation (13.4.13) of Abramowitz and Stegun (1965), this can be simplified to

$$
P = mM\left(\frac{m-D_2}{2},\frac{m}{2},z\right), \tag{7.23}
$$

and after expressing z in terms of r again, we find

$$
\frac{dC_m(r)}{dr} = 2ar^{D_2-1}\left(\frac{r^2}{4\sigma^2}\right)^{\frac{m-D_2}{2}}\frac{\Gamma(D_2/2+1)}{\Gamma(m/2)}M\left(\frac{m-D_2}{2},\frac{m}{2},-\frac{r^2}{4\sigma^2}\right).
$$
$$\tag{7.24}$$

This result coincides with the density of distances, given in Equation (7.19), derived from our expression of the Gaussian kernel correlation integral, if a is taken to be

$$
a = \phi(\Gamma(D_2/2+1))^{-1}e^{-m\tau K_2}m^{-d/2}2^{-D_2}. \tag{7.25}
$$

This identification completes the link between Smith's expression for the correlation integral in the presence of noise, given in Equation (7.2), and our expression for the Gaussian kernel correlation integral. Notice that the right hand side of Equation (7.25) does not depend on r, as expected, since both our expression and Smith's expression describe the same r-dependence. It does however depend on m. The latter is the result of the fact that Smith considered the correlation integral with m fixed, in which case a is a constant.

Equations (7.2) and (7.25) give the following expression for the correlation integral:

$$C_m(r) = \frac{\phi e^{-m_T K_2} m^{-D_2/2} 2^{-m} \sigma^{D_2-m} r^m}{\Gamma(m/2+1)} M\left(\frac{m-D_2}{2}, \frac{m}{2}+1, -\frac{r^2}{4\sigma^2}\right)$$
(7.26)

in the presence of noise when $C_m(r)$ is based on the Euclidean norm.

For fixed embedding dimension m, Equation (7.19) can be used to obtain a maximum likelihood estimate of the correlation dimension D_2, and the noise level σ. Due to the complicated expression for $\rho(r)$, the likelihood equations can not be solved analytically, and a numerical procedure for solving them should be used. Smith (1992a) showed that this methods works well for the Hénon map with noise up to noise levels of about 2%.

3.1. *Coarse-grained quantities*

For the coarse-grained correlation dimension, Equations (7.19) and (7.26) give,

$$
\begin{aligned}
D_{2,m}(r) &= \frac{d \ln C_m(r)}{d \ln r} \\
&= m \frac{M((m-D_2)/2, m/2, -r^2/(4\sigma^2))}{M((m-D_2)/2, (m+2)/2, -r^2/(4\sigma^2))} \\
&= m \frac{M(D_2/2, m/2, r^2/(4\sigma^2))}{M((D_2+2)/2, (m+2)/2, r^2/(4\sigma^2))},
\end{aligned}
$$
(7.27)

where the latter equality results from Kummer's equation (Equation (13.1.27) of Abramowitz and Stegun, 1965), and which is the same as the expression found by Schreiber (1993a). This gives the limits,

$$D_{2,m}(r) \to m, \qquad \text{for } r^2/(4\sigma^2) \to 0, \tag{7.28}$$

and

$$D_{2,m}(r) \to m \frac{\Gamma(m/2)\Gamma(D_2/2+1)}{\Gamma(D_2/2)\Gamma(m/2+1)} = D_2, \qquad \text{for } r^2/(4\sigma^2) \to \infty, \tag{7.29}$$

by Equations (13.5.5) and (13.1.4) of Abramowitz and Stegun (1965) respectively. This shows that the coarse-grained correlation dimension changes from $D_{2,m}(r) \approx m$ (for $r \ll \sigma$) to $D_{2,m}(r) \approx D_2$ (for $r \gg \sigma$).

In the previous chapter it was noted that $d_{m,n}(r)$, for m, n large and for the correlation integrals calculated with the supremum norm, is a sigmoidal function of r, which is independent of m and n for sufficiently large m and n and falls off from 1 to 0 on a scale proportional to σ. A similar result was obtained for the Gaussian kernel. For the correlation integral with the Euclidean norm we find

$$
\begin{aligned}
d_{m,n}(r) \;=\;& (D_{2m}(r) - D_{2n}(r))/(m - n) \\[2mm]
=\;& \frac{mM(D_2/2, m/2, r^2/(4\sigma^2))}{(m - n)M((D_2 + 2)/2, (m + 2)/2, r^2/(4\sigma^2))} \\[2mm]
& - \frac{nM(D_2/2, n/2, r^2/(4\sigma^2))}{(m - n)M((D_2 + 2)/2, (n + 2)/2, r^2/(4\sigma^2))},
\end{aligned}
\tag{7.30}
$$

for which Equations (7.28) and (7.29) give the limits

$$
\lim_{r/(4\sigma^2)\to 0} d_{m,n}(r) = 1
\tag{7.31}
$$

and

$$
\lim_{r/(4\sigma^2)\to\infty} d_{m,n}(r) = 0.
\tag{7.32}
$$

Although these limits are consistent with a sigmoidal function falling off from 1 to 0, they do not allow us to conclude whether $d_{m,n}(r)$ is independent of m and n for large m and n.

From Equation (7.26) the following expression for the coarse-grained entropy is found:

$$
\begin{aligned}
K_{2,m}(r) \;=\;& \frac{1}{\tau}\left(\ln C_m(r) - \ln C_{m+1}(r)\right) \\[2mm]
=\;& K_2 + \frac{1}{\tau}\ln\left(\frac{\Gamma((m + 3)/2)}{\Gamma((m + 2)/2)}\right) \\[2mm]
& + \frac{D_2}{2\tau}\ln\left(\frac{m + 1}{m}\right) - \frac{1}{2\tau}\ln\left(\frac{r^2}{4\sigma^2}\right) \\[2mm]
& + \frac{1}{\tau}\ln\left(\frac{M((D_2 + 2)/2, (m + 2)/2, r^2/(4\sigma^2))}{M((D_2 + 2)/2, (m + 3)/2, r^2/(4\sigma^2))}\right).
\end{aligned}
\tag{7.33}
$$

Using the fact that $\Gamma((m + 3)/2)/\Gamma((m + 2)/2) \sim m^{1/2}$ for large m, and that the third and last term converge to zero for large m, one finds

$$
K_{2,m}(r) \simeq K_2 - \frac{1}{\tau}\ln\left(\frac{r}{2\sqrt{m}\sigma}\right)
\tag{7.34}
$$

for $r \ll \sigma$ and large m. This expression can be seen to be similar to that found for the Gaussian kernel correlation integral (Equation (6.32)) if one identifies $h^2 m$ and r^2. The embedding dimension m enters the expression since the variance of the Gaussian kernel function scales as mh^2 whereas the variance of the Θ kernel just converges to r^2 for large m. If we define $r = r'\sqrt{m}$, for $K'_{2,m}(r') = (\ln C_m(\sqrt{m}\, r') - \ln C_{m+1}(\sqrt{m+1}\, r'))/\tau$ we obtain

$$K'_{2,m}(r') \simeq K_2 - \ln\left(\frac{r'}{2\sigma}\right) \tag{7.35}$$

instead, which is independent of m.

4. Concluding Remarks

We derived an analytic expression for the correlation integral for the Euclidean norm in the presence of Gaussian observational noise. This expression was obtained by linking the correlation integral, via a Laplace transform, to the Gaussian kernel correlation integral, for which an analytic expression was found in Chapter 6.

A maximum likelihood method for the estimation of the correlation dimension of attractors was shown by Smith (1992a) to work well in numerical simulations for attractors with small amounts of observational noise. We feel, however, that estimation of dynamical invariants based on the traditional correlation integral is rather unnatural in the presence of Guassian observational noise. The fact that a general expression with a simple analytic structure is easily obtained for Gaussian kernel correlation integrals indicates that the Gaussian kernel correlation integrals are a natural tool for examining the structure of attractors with Gaussian noise. Indeed, while the maximum likelihood method for the estimation of the correlation dimension appears to break down at noise levels as low as a few percent for the Hénon map (Smith, 1992a), the results of the previous chapter indicate that nonlinear fits of the Gaussian kernel correlation integral can be used for the estimation of the correlation dimension, the noise level and the correlation entropy for noise levels up to 20%.

Oltmans and Verheijen (1997) derived an expression for the correlation integral in the presence of Gaussian noise by direct analytic calculation under the assumption that the density of distances on the attractor scales as r^{D_2-1} for all $r > 0$. In fact, in the derivation of the expression for the Gaussian kernel correlation integral in Chapter 6 this was assumed implicitly. Schreiber (1993a) has examined the correlation integral in the presence of Gaussian noise for the supremum norm. This approach allows for the estimation of the noise level σ by fitting the correlation integral for various values of the embedding dimension m. Schreiber (1997) compared a number of different approaches, and found that all approaches give qualitatively similar curves of the coarse-grained dimension as a function of the bandwidth.

CHAPTER 8

SPIRAL WAVE TIP DYNAMICS

In this chapter we analyze time series of meandering spiral waves in a two-dimensional model of excitable media. A regular as well as a more irregular looking time series of the spiral tip position are studied. In the irregular case the tip position appears to be non-stationary on the time scale considered (of the order of 100 tip rotations). It is not possible to give meaningful estimates of the correlation dimension D_2 and correlation entropy K_2 in this case. However, low-dimensional behavior is found in the spiral tip velocity for the irregular series. Results obtained with the noisy attractor model suggest that the velocity time series of the irregular looking tip trajectory is two-frequency quasi-periodic rather than low-dimensional chaotic.

1. Introduction

In the last decades, the theory of nonlinear dynamical systems has attracted attention due to the fact that many systems which are seemingly random turned out to be low-dimensional chaotic. Even a very complex system like a normally functioning heart is stated to exhibit behavior that can be described by the interaction of just a few state variables (Babloyantz and Destexhe, 1988). On the mathematical side progress is being made, and for some classes of partial differential equations (Debussche and Marion, 1992) it has been proven that the dynamics in state space are confined to a finite dimensional subspace, so that the system can be considered to be finite dimensional. The dimension, although finite, may however well exceed the empirical upper limit of about 5 for dimensions estimated from a time series.

A wealth of methods have been developed to extract dynamical information from scalar experimental time series, which all in one way or another are based on the reconstruction theorem of Takens (1981). In principle, for spatio-temporal systems such as the heart, measurements at one point in space should contain the necessary information to reconstruct the dynamics, provided that the system is finite and the various parts are sufficiently coupled (Grassberger *et al.*, 1991).

In this chapter we will analyze time series obtained from a two-dimensional model of excitable tissue, exhibiting spiral wave activity. Spiral waves are observed in various reaction-diffusion systems, such as the Belousov-Zhabotinsky reaction (Aliev and Rovinsky, 1992; Fife, 1985; Keener and Tyson, 1986), aggregation of Dictyostelium Discoideum amoebae (Tyson *et al.*, 1989), and heart tissue (Allessie *et al.*, 1977; Davidenko *et al.*, 1991; Davidenko *et al.*, 1992). Several authors have conjectured that some cardiac arrhythmias may be initiated by spiral waves; the break-up of spiral waves in models of heart tissue has been studied by Karma (1993), Panfilov and Holden (1991) and by Panfilov and Hogeweg (1993).

If the tip of a spiral wave does not exhibit steady rotation around a point in the plane, the spiral wave is said to be meandering. The tip in so-called floral meander follows a flower-like pattern and the orbit traces out an annular region of the plane. More irregular types of meander, called hyper-meander, have been described by Jahnke and Winfree (1991) and Courtemanche and Winfree (1991). Hyper-meander generally occurs in parameter regions of reaction-diffusion models with a large ratio of the speeds of the excitation and recovery variables, together with a relatively large excitability threshold. Here we examine whether hyper-meander is deterministic and possibly chaotic as suggested by Jahnke and Winfree (1991).

The meandering spiral waves examined here are generated with the model introduced by Barkley (1991). We are interested in the behavior of spirals in media in which the spiral tip is only marginally influenced by the size of the system. In our numerical experiments this goal was met by taking the system size sufficiently large, and choosing initial conditions so that the spiral tip did not come too close to the boundaries. In contrast, a study of the interaction of spiral waves with the boundary in a small medium was performed by Davidov and Zykov (1993).

2. Wave Tip Meandering

The equations of the excitable media model introduced by Barkley (1991) are

$$\frac{\partial u}{\partial t} = \frac{1}{\epsilon}u(1-u)(u - \frac{v+b}{a}) + D\Delta u$$
$$\frac{\partial v}{\partial t} = u - v,$$

(8.1)

where u is the fast excitation variable and v the slow recovery variable. Both variables are a function of space and time, and the diffusion term in the equation for the fast excitation variable u accounts for a coupling of the dynamics at different position in the media. A typical excitation of the local dynamics of this model is shown in Figure 3.5. The parameter ϵ, which typically is much smaller than 1, is a measure of the relative speed of the slow recovery variable v with respect to the speed of the fast excitation variable u. The details of the local dynamics are determined by a and b, D is the diffusion constant and Δ denotes the Laplace operator. For completeness, it should be noted that the variables u and v and the model parameters a, b and ϵ are dimensionless. If needed, the model can always be rescaled to physical units by using references quantities such as wavelength, rotation time of a spiral wave, and the plane wave velocity.

The model is considered on an $L \times L$ square part of the two-dimensional plane and integrated numerically using a two-dimensional array of $n \times n$ cells imposing no-flux (reflecting walls) boundary conditions, with the implicit integration scheme proposed by Barkley (1991). Barkley also described a method for improving numerical efficiency at the expense of accuracy, which is not used here. The convergence of the numerical scheme was examined and we confirmed that the tip trajectories converge quadratically in the time discretization as well as in the space discretization. We fixed the size parameter at $L = 20$ and the diffusion parameter at $D = 1$ for all calculations. We found good convergence together with a balance between the errors due to space and time discretization for $n = 190$ and a time step $dt = 10^{-3}$ t.u. (time units). The spirals were initialized by inducing a target wave on the middle left side $(x, y) = (0, \frac{1}{2}L)$ and integrating the model equations until the wave front reached the point $(x, y) = (\frac{3}{4}L, \frac{1}{2}L)$. At that time we set $t = 0$ and the state on the upper half of the plane was set to the stable equilibrium point of the local dynamics $(u, v) = (0, 0)$ after which the loose end of the front developed a meandering spiral tip.

Following Barkley *et al.* (1990), we define the x and y position of the spiral tip to be the intersection of the lines in the plane where $u = \frac{1}{2}$ and $f(u, v) = 0$ with

$$f(u, v) = \frac{1}{\epsilon} u(1 - u)(u - \frac{v + b}{a}).$$ (8.2)

Other definitions of tip positions can be found in the literature, such as that proposed by Lugosi (1989), but we found the definition in Equation (8.2) to be relatively insensitive to the effects of spatial discretization. The essential aspect of any tip definition is that one and only one point in the plane satisfies all conditions in the presence of a single spiral wave.

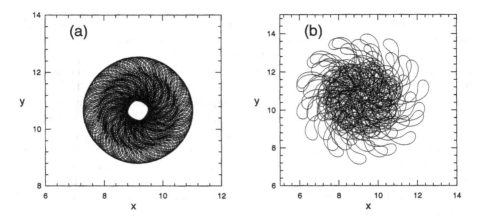

Figure 8.1: *Tip trajectories for a = 0.3, b = 0.01 and ε = 1/200 (a) and for a = 0.25, b = 0.001 and ε = 1/200 (b).*

At equal time intervals $\Delta t = 0.04$ t.u., we calculated and stored the tip position $(x_{\text{tip}}, y_{\text{tip}})$, which was determined numerically by linear interpolation of the (u, v) field within the square cells of the grid and solving the intersections of the curves on which $u = \frac{1}{2}$ and $f(u, v) = 0$. We stored 5×10^3 tip positions from $t = 100.04$ to $t = 300.00$. The time up to $t = 100$ corresponds to about 50 spiral rotations. After this period transient effects had decayed below the level of the numerical accuracy.

We examine two different cases of meander in Barkley's model: a regular looking motion and an erratic looking one. Figure 8.1(a) shows the trajectory followed by the tip for the regular case. The model parameters for this series are $a = 0.3$, $b = 0.01$ and $\epsilon = 1/200$. The tip position appears to exhibit two-frequency quasi-periodic motion and traces out an annular region of the plane. The small four-fold deviation from a circle which can be observed in the annulus in Figure 8.1(a) is a small boundary effect. Figure 8.1(b) shows the tip trajectory obtained in a simulation with $a = 0.25$, $b = 0.001$ and $\epsilon = 1/200$. The corresponding tip trajectory has a much more irregular appearance. The two time series of the x-coordinate of the tip position are shown in Figure 8.2.

We will examine whether these time series can be characterized by the dynamical invariants D_2, the correlation dimension, and K_2, the correlation entropy (see Equation (2.22) for their definitions). The $x_{\text{tip}}(t)$ time series are mapped into an m-dimensional state space by constructing delay vectors according to $\boldsymbol{X}(t) = (x_{\text{tip}}(t), x_{\text{tip}}(t + \tau), \ldots, x_{\text{tip}}(t + (m - 1)\tau))$. The reconstruction theorem (Takens, 1981) states that this mapping generically gives a

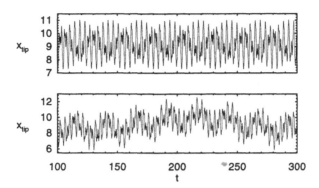

Figure 8.2: *Time series of the x-coordinate for $a = 0.3$, $b = 0.01$ and $\epsilon = 1/200$ (upper panel) and for $a = 0.25$, $b = 0.001$ and $\epsilon = 1/200$ (lower panel).*

faithful representation of the original attractor provided that the embedding dimension m is sufficiently large. The choice of the embedding parameters m and τ is not important in the reconstruction theorem, but in practice where we do not have an infinite number of noise-free measurements, it is essential. As suggested by Fraser and Swinney (1986) we set the delay τ equal to the time shift where the first minimum in the mutual information occurs. The correlation dimension D_2 is estimated from the local slope of the correlation integral by the Grassberger-Procaccia method with a Theiler correction, T, which was set to twice the delay time τ throughout.

The regular time series is embedded with $\tau = 0.64$ t.u. and the plot of the correlation integral $C_2(r)$ versus the radius r is shown in the lower panel of Figure 8.3 for $m = 2$ to $m = 20$. The local slope $d \log(C_2(r))/d \log(r)$ is shown in the upper panel. Figure 8.4 shows the estimated correlation dimension and correlation entropy as a function of the embedding dimension m. The estimated D_2 saturates at about 2.1, which is only slightly above the value 2 which we would expect for a regular motion on a two-torus. The correlation entropy K_2 converges to zero, a necessary condition for regular motion. These results, together with the appearance of Figure 8.1(a) as a projected torus, strongly suggest that the tip exhibits two-frequency quasi-periodic behaviour. The erratic $x_{\mathrm{tip}}(t)$ time series of Figure 8.1(b) has a first local minimum in its mutual information function at a shift of 1.0 t.u. The correlation integral and the local slope estimated with this delay time for $m = 2$ to 20 are shown in Figure 8.5. A small scaling region can be observed between $r = 0.15$ and $r = 0.30$, but these distances are very large the scaling region too small for a reliable dimension estimate.

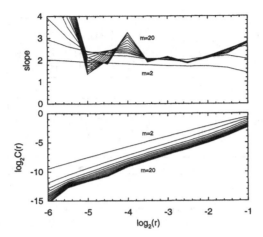

Figure 8.3: $\log_2(C(r))$ vs. $\log_2(r)$ for the regular time series for values of the embedding dimension m ranging from 2 to 20 (lower panel) and the local slope (upper panel).

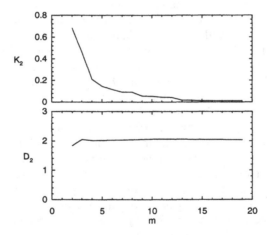

Figure 8.4: Estimated correlation dimension D_2 (lower panel) and correlation entropy K_2 (upper panel) as a function of embedding dimension m for the regular tip time series.

Recurrence plots (Eckmann *et al.*, 1987) are useful for examining the stationarity of a time series. If at a certain time t_2 the state in the reconstructed

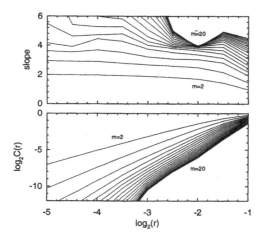

Figure 8.5: $\log_2(C(r))$ vs. $\log_2(r)$ for the erratic time series for values of the embedding dimension m ranging from 2 to 20 (lower panel) and local slope (upper panel).

state space is closer than some specified minimum distance ϵ to a previously visited state, the state is said to have recurred to that previous state (at resolution ϵ). A recurrence plot has two time axes and shows a dot for each pair (t_1, t_2) of times for which the two corresponding states are closer than ϵ. If a time series is stationary, the recurrence plot will show an approximately uniform density of recurrences as a function of the time difference $t_1 - t_2$. However, if the time series has a trend or another type of non-stationarity, that is, if its behaviour changes over time, the regions of the state space visited will change in time. The result will be that there are relatively few recurrences far from the diagonal in the recurrence plot, i.e. for large values of $|t_1 - t_2|$. Also, if there are only recurrences near $t_1 = t_2$ and for values of $|t_1 - t_2|$ that are of the order of the total length of the time series, the time series can be considered non-stationary for practical purposes. As an example, consider a stationary time series superimposed on a slowly varying periodic time series. On large time scales, one can consider this time series stationary, but on small time scales it will behave as a non-stationary time series.

Figure 8.6 shows recurrence plots of the two time series obtained with $\epsilon = 0.1$ at embedding dimension $m = 10$. The histograms in the upper panels in this figures show the number of recurrences as a function of the distance from the diagonal. For the regular time series there are many recurrences and

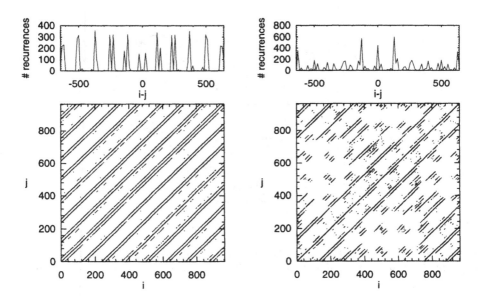

Figure 8.6: *Recurrence plots (lower panels) for the regular (left) and irregular (right) looking time series. The time series were sampled down by a factor 5, resulting in a sample time of 0.2 t.u. and embedded with a delay of 4 sample times in embedding dimension $m = 10$. The upper panels show the number of recurrences as a function of the distance from the diagonal line.*

the number of recurrences shows no signs of dependence from the distance to the diagonal line, suggesting that the regular time series is stationary. For the irregular time series, structures can be seen in the recurrence plot that extend over a time scale that is nearly as large as the entire period of measurement. The histogram has a number of peaks near the center and a few smaller peaks near both sides of the histogram. Although these peaks represent recurrences, they occur on time intervals comparable to that of the entire time series, which indicates that the erratic time series is not stationary, at least not on the time scale over which the measurements are performed.

A well-known method to remove non-stationarities is that of differencing the time series, i.e. constructing the time series $x_{\text{tip}}(t + s) - x_{\text{tip}}(t)$. The delay s can be used to optimize the signal to noise ratio of the series. Assuming that $x_{\text{tip}}(t)$ consists of a clean signal plus independent, identically distributed noise, the variance of the noise in the differenced signal will not depend on the delay, while the variance of the signal has a maximum if s is near the minimum of the mutual information.

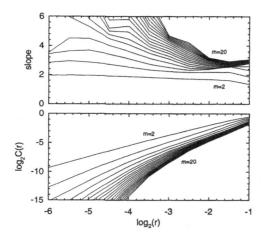

Figure 8.7: $\log_2(C(r))$ vs. $\log_2(r)$ for the differenced erratic time series for values of the embedding dimension m ranging from 2 to 20 (lower panel) and the local slope (upper panel).

Using $s = 1.0$ t.u., equal to the value of the first local minimum in the mutual information function, the erratic time series gave an apparently stationary time series already after the first time of differencing. The logarithmic plot of $C_2(r)$ is shown in Figure 8.7 together with the local slope. A clear scaling region cannot be identified in this figure. There is a relatively small region where the local slope is nearly constant with a value of about 3, but this occurs for large values of r. We consider the corresponding values of r too large and the scaling region too small for reliable estimation of the correlation dimension and the correlation entropy. This result does not change after further differencing.

If the tip position is not low-dimensional, as our results suggest, we also expect to find low-dimensional behaviour neither for a bounded time series like that of $u(x, y)$ or $v(x, y)$ measured at a fixed position in the plane, nor for an average over a certain region in the plane. In order to check this, we use a time series based on the average values of u in the lower half of the plane. This average is bounded between zero and one, which excludes arbitrary large fluctuations of the amplitude. No scaling region is present in the correlation integral of this time series and these results remain similar after differencing the time series.

3. Measurement Function

Symmetry considerations enable the construction of a measurement function of the system that simplifies the characterization of the behaviour of the (u, v) field near the tip.

In Figure 8.1 there appears to be much more recurrence in the observed variables x_{tip} and y_{tip} for the regular time series than for the erratic one. In the regular case there are close recurrences and the trajectories closely follow near trajectories. In the erratic case, close recurrences to previous states are less numerous, as already suggested by the low values of the correlation integrals for small values of r. From the regular tip trajectory in Figure 8.1(a) it is clear that there are many trajectory segments that have a nearly identical shape. The main difference between these segments is that they have another starting point in the plane and another angle under which the trajectory leaves the starting point. A close inspection of the erratic time series shows similar features. This suggests that segments of the trajectory modulo rotations and translations show more recurrences than observed in the original trajectory.

The idea now is to construct a measurement function that is independent of rotations and translations, and can be used to characterize the tip dynamics modulo rotations and translations. A function which possesses these properties for example is the tip velocity $[(dx_{\text{tip}}/dt)^2 + (dy_{\text{tip}}/dt)^2]^{1/2}$, because it is the same in all translated and rotated frames in the plane. In practice, where we have a finite sample time, it is more convenient to use the square of the Euclidean distance travelled by the tip between times t and $t + r$,

$$h(t) = [(x_{\text{tip}}(t + r) - x_{\text{tip}}(t))^2 + (y_{\text{tip}}(t + r) - y_{\text{tip}}(t))^2]^{1/2}. \qquad (8.3)$$

It can be easily checked that this function has the desired symmetry properties, that is, it is also a function that does not depend on rotations and translations. The construction of the time series $h(t)$ may be considered as a generalization of differencing (as done for scalar time series in the previous section) to vector time series.

There are situations where the dimension of $h(t)$ is low, like in the steady rotation in the FitzHugh-Nagumo model described by Karma (1990). Because the angular velocity $\omega(t)$ and tangential velocity $s(t)$ of the tip trajectory generally will change with the local shape of the spiral wave rotating in the medium, a constant $\omega(t)$ and $s(t)$ imply a constant shape of the piece of trajectory, corresponding to a fixed point solution of $h(t)$ which has dimension zero.

The time series $h(t)$ for the regular time series was generated with $r = 0.64$ t.u., the time shift for which the mutual information function has its first minimum. We checked that the same delay was found when the y_{tip} signals were used. Figure 8.8(a) shows a phase portrait of $h(t)$. It resembles a limit cycle with a small amount of noise. We subscribe this effect mainly to the influence of the boundary, because in the x_{tip} time series the noise level corresponds to the discretization noise level but in $h(t)$ it is much higher. As mentioned before, Figure 8.1(a) already suggests that there are small boundary effects since the inner of the annulus shows a small four-fold deviation from a circle.

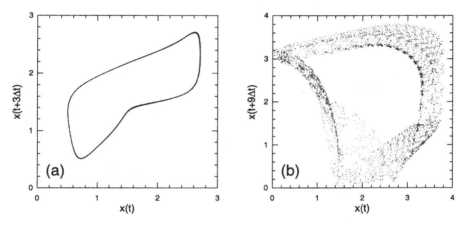

Figure 8.8: *Phase portrait of $h(t)$ for the regular time series with a delay of 0.12 t.u. (a) and the erratic time series with a delay of 0.36 t.u. (b).*

A description of the spiral tip position in terms of a composition of two circular rotations with two different frequencies has been proposed by several authors using other models (Lugosi, 1989). The composition of two rotations with amplitudes r_1 and r_2, and angular frequencies ω_1 and ω_2 respectively gives,

$$
\begin{aligned}
x_{\text{tip}} &= r_1 \cos(\omega_1 t + \varphi_1) + r_2 \cos(\omega_2 t + \varphi_2) \\
y_{\text{tip}} &= r_1 \sin(\omega_1 t + \varphi_1) + r_2 \sin(\omega_2 t + \varphi_2).
\end{aligned}
\tag{8.4}
$$

Under the assumption of composite rotation $h(t)$ becomes

$$
\begin{aligned}
h(t) = \ & 4r_1^2 \sin^2(\omega_1 r/2) + 4r_2^2 \sin^2(\omega_2 r/2) \\
& + 8r_1 r_2 \cos((\omega_1 - \omega_2)(t + r/2) + \varphi_1 - \varphi_2) \\
& \times \sin(\omega_1 r/2) \sin(\omega_2 r/2),
\end{aligned}
\tag{8.5}
$$

which by the occurrence of t only through the cosine should be time reversible
and have a phase portrait that is symmetric with respect to reflection in the
diagonal line. Figure 8.8(a) lacks this symmetry, suggesting that the meander
does not allow a description in terms of composite rotation. In other words,
there is no uniformly rotating reference frame in which the tip travels uniformly
along a circle. This has also been observed by Barkley *et al.* (1990) in a slightly
different model. Figure 8.8(b) shows a phase portrait of $h(t)$ for the erratic
time series and it clearly shows more structure and close recurrences than the
original tip trajectory in Figure 8.1(b).

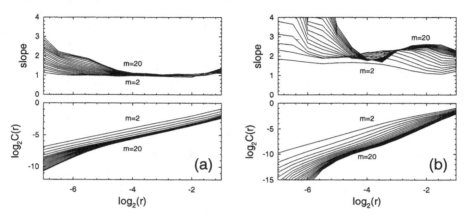

Figure 8.9: *Correlation integrals* $\log_2(C(r))$ *vs.* $\log_2(r)$ *(lower panels) and local
slope (upper panels) for* $h(t)$ *of the regular series (a) and* $h(t)$ *of the erratic time
series (b).*

Figure 8.9 shows the correlation integrals for $h(t)$ of both time series. The
corresponding estimates of the correlation dimension and correlation entropy
are shown in Figure 8.10. The estimated correlation dimension D_2 of $h(t)$ for
the regular time series is slightly larger than 1.0 and the correlation entropy
K_2 appears to converge to 0 nats/t.u. The value of D_2 estimated from the
scaling region of the correlation integral of $h(t)$ for the erratic series converges
to about 2 as a function of embedding dimension. The correlation entropy K_2
appears to converge to a value of about 0.13 nats/t.u. Although the entropy
does not converge to zero, we can not exclude the possibility that this series
is two-frequency quasi-periodic; the dimension is close to 2 while the positive
entropy may be the result of integration noise.

Note that the difference between the dimension of the regular x_{tip} series and
of its $h(t)$ series is about 1. This is not coincidental, but the result of the lack

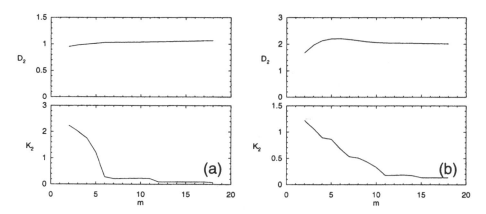

Figure 8.10: *Estimated correlation dimension and correlation entropy for $h(t)$ of the regular time series (a) and the erratic time series (b).*

of rotational symmetry invariance present in the x_{tip} measurement function. This measurement function contains a dependence on the angle made by the piece of trajectory with the x-axis. Since this angle can take all possible values between 0 and 2π, independently of the value of $h(t)$, there will be a dimension in the reconstructed state space which represents this angle.

It remains to be discussed why the correlation integral of the differenced erratic x_{tip} series did not show scaling (see Figure 8.7). Using the argument that we used above for the regular tip, we would expect a dimension which is 1 larger than the dimension of the $h(t)$ time series, for which we found $D_2 \approx 2$. A possible explanation is that the attractor has a folded structure as a result of the angle dependence, causing an effective increase of the noise level, so that no scaling can be observed at small length scales. Near the largest values of r, the correlation integrals in Figure 8.7 display a very small scaling region with a slope of about 3, which, in the perspective given above, may indeed be a residual of a much larger scaling region with a slope equal to 3.

3.1. Fitting the Noisy Attractor Model

We applied the Gaussian kernel correlation integral method (see Chapter 6) to the time series obtained with measurement function h for the irregular looking tip trajectory, in order to see whether a description of this time series in terms of a noisy deterministic time series is reasonable. The dynamical invariants and noise level estimated from the Gaussian kernel correlation integral

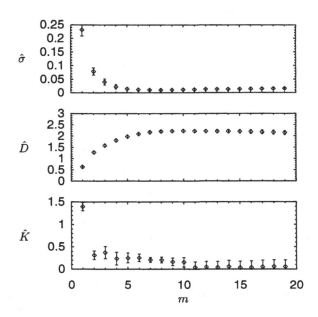

Figure 8.11: *Estimated values $\widehat{\sigma}$, $\widehat{D}_2$ and $\widehat{K}_2$ for the time series $h(t)$ obtained from the irregularly looking tip trajectory. The bars denote the estimated 95% confidence interval (2 standard errors).*

are shown in Figure 8.11. We used a delay time τ of 0.36 time units for the reconstruction, a Theiler correction $T = 2\tau$ and a fit of $T_m(h)$ in the bandwidth region $[0.0001, 0.0625]$. The results are highly suggestive of a two-torus with noise; the estimated dimension is close to two, whereas the estimated value of the correlation entropy converges to zero. The noise level is found to be about 0.01. These findings suggest furthermore that the noise model for the Gaussian kernel correlation integral, although designed for situations with observational noise, may be useful in the analysis of time series obtained from dynamical systems with dynamical noise (or numerical integration noise).

4. Model of Meandering

In this section a simple model of meandering is proposed and used to generate tip trajectories that can be compared to the trajectories from Barkley's model. Using this model, a possible explanation can be given for the non-stationarity of the erratic tip time series.

We expect that the low values of the correlation dimension that we found for $h(t)$ are somehow related to low-dimensional behavior of the (u, v) field near the tip, because the dynamics of the tip position in infinitesimal time intervals should be determined completely by the fields near the tip. This is in accordance with the observation of Lugosi (1989) that the shape of the spiral wave near the tip in the model of Zykov changes during floral meander, but is the same whenever it is in the same stage of the meander. We therefore conjecture that we have reconstructed the dynamics of the (u, v) field near the tip using the measurement function $h(t)$. Of course, this can only hold under the assumption that the field near the tip behaves low-dimensional, for otherwise the reconstruction theorem does not apply.

Assuming that $h(t)$ is low-dimensional we may write the dynamics for a delay vector $x(t) = (h(t), h(t + \tau), \ldots, h(t + (m - 1)\tau))$ in the reconstructed state space as,

$$\frac{dx}{dt} = F(x), \tag{8.6}$$

where F is some (unknown) vector field. Thus, the vector $x(t)$, which represents the state of the field near the spiral tip, is assumed to evolve according to an autonomous ordinary differential equation with a finite number of variables. Now we would like to obtain a description of how x_{tip} and y_{tip} depend on $x(t)$. The evolution of orientations and translations is described by the variables $x_{\text{tip}}, y_{\text{tip}}$ and the angle α of the tip velocity vector with the x-axis. The time derivatives of x_{tip} and y_{tip} define the tip velocity $s(t)$, and the time derivative of α is the angular velocity $\omega(t)$. We have $d\alpha/dt = \omega(t)$ and, $dx_{\text{tip}}/dt = s(t)\cos(\alpha)$ and $dy_{\text{tip}}/dt = s(t)\sin(\alpha)$.

Like $h(t)$, the function $s(t)$ is a proper measurement function for the reconstruction of the tip dynamics. This implies the existence of a one-to-one map from the attractor in the state space reconstructed with $h(t)$ to the attractor in the state space reconstructed with $s(t)$, so that we may write $s(t) = s(x(t))$. An analogous argument shows that $\omega(t) = \omega(x(t))$. Taking together, we have

$$\begin{aligned}
\frac{d\alpha}{dt} &= \omega(x) \\
\frac{dx_{\text{tip}}}{dt} &= s(x)\cos(\alpha) \\
\frac{dy_{\text{tip}}}{dt} &= s(x)\sin(\alpha),
\end{aligned} \tag{8.7}$$

describing how x_{tip} and y_{tip} are driven by the variable x which represents the state of the local fields.

The form of the functions $\omega(x)$ and $s(x)$ are unknown, but we will use the model in Equations (8.6) and (8.7) to deduce some qualitative features of the behavior of the tip time series. The main goal at this stage is to infer some general relations between the asymptotic dynamical behavior of $h(t)$ and the behavior of the tip position. If x is attracted to a steady state, asymptotically $s(x) = constant$ and $\omega(x) = constant$, corresponding to a rigidly rotating spiral. If, however, the asymptotic behavior of $h(t)$ is periodic, after one period, p say, of $h(t)$ we again have the same state x, but we expect that due to Equation (8.7) a translation of the spiral has occurred, together with a rotation. This procedure repeats itself and if the rotation $\alpha(t+p)-\alpha(t)$ is incommensurate with 2π the tip exhibits two-frequency quasi-periodic motion and its trajectory traces out an annulus in the plane.

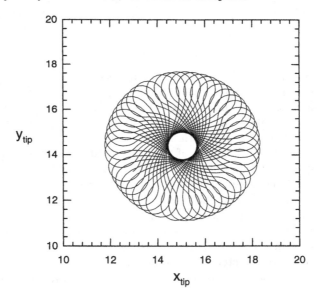

Figure 8.12: *Model tip trajectory where the driving term x in Equation (8.7) was generated with the Van der Pol equations.*

As an example we integrated the Van der Pol equations (see e.g. Broer and Dumortier, 1991)

$$\frac{\mathrm{d}x_1}{\mathrm{d}t} = x_2$$
$$\frac{\mathrm{d}x_2}{\mathrm{d}t} = 2.0(1.0 - x_1^2)x_2 - x_1,$$

(8.8)

and used time series with sample time $\Delta t = 0.2$ to obtain a model time series for which x is periodic, together with Equation (8.7), with a linear dependence of ω and s on x:

$$\omega(x) = 0.5x_1 + 0.5$$
$$s(x) = 0.25x_2 + 1.0.$$

(8.9)

The resulting tip trajectory, shown in Figure 8.12, resembles the projection of a two-torus (cf. Figure 8.1(a)).

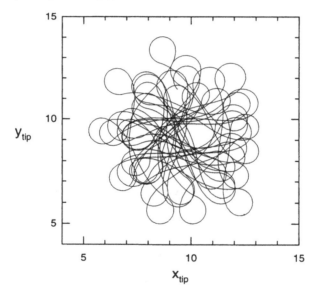

Figure 8.13: *Model tip trajectory obtained with Equation (8.7) using a quasi-periodic driving term x from a linear combination of two Van der Pol models with incommensurate time scales.*

A two-frequency quasi-periodic driving force x can be obtained by using $x = y + z/10$, where y and z are both generated with the Van der Pol equations (8.8). The time scale of z was modified by dividing the right hand side of the Van der Pol equations by the irrational number $\nu = (1 + \sqrt{5})/2$. The series x was used together with Equations (8.7) and (8.9) to generate a model tip trajectory. This trajectory is shown in Figure 8.13. The trajectory bears some resemblance to Figure 8.1(b).

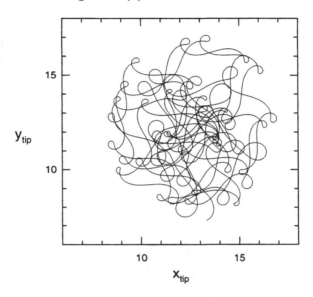

Figure 8.14: *Model tip trajectory generated by Equation (8.7) where the driving term x was generated by the Rössler equations.*

Imposing chaotic behavior on x by using the Rössler equations

$$
\begin{aligned}
\frac{dx_1}{dt} &= -x_2 - x_3 \\
\frac{dx_2}{dt} &= x_1 + 0.15x_2 \\
\frac{dx_3}{dt} &= 0.2 + x_3(x_1 - 10.0),
\end{aligned}
\tag{8.10}
$$

and then integrating Equation (8.7) with dependence

$$
\begin{aligned}
\omega(x) &= 0.15x_1 + 0.8 \\
s(x) &= 0.05x_2 + 0.85
\end{aligned}
\tag{8.11}
$$

gives the tip position shown in Figure 8.14. Also here the sample period Δt is 0.2. The time series obtained with this chaotic driving term is non-stationary.

Longer time series generated with the model exhibit features of random walks. This might be an example of deterministic diffusion, caused by sensitive dependence on initial conditions of the driving term $x(t)$. Deterministic diffusion of spiral wave tip positions in excitable media has been studied by Biktashev and Holden (1998).

5. Summary and Discussion

We have analyzed two tip trajectories in Barkley's model: a regular trajectory and an irregular looking tip trajectory. The regular tip trajectory showed various characteristics of a two-frequency quasi-periodic time series. These results were confirmed by a nonlinear time series analysis in which the regular tip trajectory was found to be consistent with a regular motion on a two-torus. For the irregular tip trajectory, we were not able to identify a clear scaling region in the correlation integral, which might be the result of the apparent non-stationarity of the time series on the time scale considered. Similar results were found using a bounded time series obtained by averaging the field over a region of the plane.

Both time series showed a scaling region upon using a specific measurement function which is invariant under translations and rotations of the activation pattern in the two-dimensional plane: the distance traveled by the tip in a given time interval. For the regular looking time series we found a correlation dimension of about 1, and about 2 for the irregular looking time series. The correlation entropy for the regular time series was found to be practically zero and to have a small positive value for the irregular looking tip time series. It is not clear whether the latter entropy is statistically significantly larger than zero or the result of numerical integration noise. This leaves open two possibilities for the irregular looking tip trajectory; it is either governed by a chaotic tip velocity with a correlation dimension of about 2, or a two-frequency quasi-periodic tip velocity.

A simple model for the tip velocity provides the following picture. The evolution of the tip position originates from the local (u, v) field. This local field interacts with the surrounding field and this interaction can give rise to low-dimensional behavior, resulting in a low-dimensional tip velocity. Let us next describe the consequences of these observations for the two possible explanations for the irregular looking tip time series.

As a consequence of sensitive dependence on initial conditions, a chaotic tip velocity would cause non-stationarity of the tip position through 'deterministic diffusion', which is clearly not low-dimensional chaotic. As a result, also bounded time series obtained by averaging the field over a region of the plane will not behave low-dimensionally chaotic. There thus is a transition from quasi-periodic to non-stationary behavior of the tip position if the behavior of the local field changes from regular to chaotic. If the local (u, v)-fields near the tip are chaotic, the tip velocity is also chaotic, giving rise to non-stationarity, so that low-dimensional behavior will not be found for the tip time series of hyper-meandering spirals in this model. Upon the use of special (translation and rotation invariant) measurement functions the low-dimensional chaotic behavior of the tip velocity can be identified.

A two-frequency quasi-periodic tip velocity will formally give a stationary tip position time series, as confirmed by numerical integration of the tip model. The observed non-stationarity of the tip position of the irregular looking tip time series could be just the result of the time scale on which we observed the tip trajectory, that is, we would observe a stationary tip time series only on time scales much longer than those covered by our numerical calculations.

Clearly, both a chaotic and a quasi-periodic tip velocity are consistent with our observations. Nevertheless, we are inclined to concluding that the irregular looking tip time series is, in fact, quasi-periodic. A strong argument in favor of the scenario of a quasi-periodic tip are the structure observed in the phase portrait of the tip velocity time series in Figure 8.8(b), which shows a very regular structure, confined to a distorted image of a torus, typical for quasi-periodic time series. This is consistent with the results obtained using the noise model of chapter 6 which suggests that the erratic looking tip velocity time series has a dimension close to 2 and a noise level of about 1%. Much longer time series will have to be used in order to determine whether the erratic looking tip trajectory is non-stationary or just stationary on a much longer time scale. To prevent numerical noise from accumulating in such a long time series, the calculations will have to be performed with a much higher numerical accuracy. The size of the medium probably will have to be taken considerably larger to further suppress possible influences of the boundaries on the long-term behavior of the tip.

In a sense, this type of spatio-temporal behavior is on the limit of what can be analyzed with the standard methods in nonlinear time series analysis. The model equations describe two time and space dependent fields in two-dimensional space. The large number of variables playing a role in the model in this case collectively show low-dimensional behavior. For excitable media with one meandering spiral waves there is only one 'organizing center' in the model; the core of the single rotating spiral. The situation will be much more complex in situations where the spiral wave breaks up into various parts and gives rise to several interacting rotating spirals, which themselves may break up again. It will then be difficult to follow a single spiral wave tip position because spiral wave tips can be absorbed by other spirals and new tips can be generated as a result of the interaction of already existing spirals. In these situations one would speak of spatio-temporal chaos. The dynamics of the system as a whole is then expected to have a dimension which is too high for chaos analysis, as the dimension and entropy of spatio-temporal chaotic systems are extensive quantities, that is, they grow linearly with the system size.

Recently Plesser and Müller (1995) analyzed periodograms of the tip trajectories in the Oregonator model. The most complex situations could be described well by 4 'structural' frequencies, to which all other frequencies in the Fourier transform were related. The interplay of these structural frequencies generated trajectories of high complexity, but no evidence of chaotic behavior was found. However, we remark that a finite time power spectrum in principle cannot provide an unambiguous distinction between a power spectrum generated by a chaotic system and a line power spectrum associated with quasi-periodic motion (Dumont and Brumer, 1988).

We note that our results may depend critically on the details of the model such as the presence or absence of a diffusion term for the recovery variable. Therefore, they may fail to hold for other reaction-diffusion models like those describing the Belousov-Zhabotinsky reaction. The latter is usually modeled with a diffusive recovery variable.

CHAPTER 9

SPATIO-TEMPORAL CHAOS: A SOLVABLE MODEL

This chapter describes a solvable coupled map lattice model exhibiting spatio-temporal chaos. Exact expressions are obtained for the spectrum of Lyapunov exponents as a function of the model parameters. Although the model has spatio-temporal structure, local time series measured at a single lattice site are shown to be independent and identically distributed, according to the same distribution for several values of the model parameters. The same result is obtained for the spatial series measured at a fixed time. In these cases, the information dimension density is 1, but because the information entropy density depends on the model parameters, the model is an example of a system for which the information entropy density can be obtained neither from a time series measured at a single lattice site nor from a spatial series measured at a fixed time. We conclude that in studying only a time series or a spatial series without any knowledge of the system, one could be easily led into thinking that there is no spatio-temporal structure. For a full characterization of the system, structure in time and space will have to be considered simultaneously.

1. Introduction

Low-dimensional deterministic time series can be analyzed by the existing chaos methods, but as indicated in Chapter 3, time series from spatially extended systems are yet poorly understood and no standard methods for their analysis are available. A spatially extended system consists of an infinite number of interacting state variables, arranged in a given spatial configuration extending infinitely far in all directions. The simplest example is a one-dimensional spatial array of state variables extending from minus to plus infinity. The state variables evolve in time according to a deterministic evolution law. Time is taken to be discrete, and the next state at a lattice site depends on the present state and on the states of the two nearest neighbors. As a result of the nearest neighbor interaction, local time series consisting of the consecutive states at a single lattice site often are not fully deterministic; the neighboring sites act as a noise source on the local state.

For low-dimensional dynamical systems, some results concerning noise have been obtained recently. If the noise is purely observational, i.e. the noise is superimposed upon a clean low-dimensional time series, noise reduction algorithms can be applied in order to recover the underlying deterministic time series. A survey of noise reduction algorithms can be found in Kostelich and Schreiber (1993). Another approach consists of estimating the dynamical invariants of a time series directly from the empirical reconstruction measure of the noisy time series. This can be achieved for example by constructing a model of the correlation integral for a time series in the presence of noise in terms of the dynamical invariants and the noise level. Schouten *et al.* (1994a) considered bounded observational noise while the model for the correlation integral in the presence of Gaussian observational noise (Smith, 1992b; Oltmans and Verheijen, 1997) can be simplified by the use of a Gaussian kernel analog of the correlation integral (see Chapter 6). Note that these algorithms may give misleading results for systems with dynamical noise, that is, dynamical systems in which the noise is an intrinsic part of the dynamics. In that case, the noise acts on the state of the dynamical system, thereby influencing the evolution. The problem of separating the clean signal from the noise is then ill-posed as argued by Grassberger *et al.* (1991).

At present little is known concerning the analysis of data obtained from spatially extended dynamical systems. In some respects the time series obtained from these models are similar to noisy time series (Mayer-Kress and Kaneko, 1989; Torcini *et al.*, 1991; Tsimring, 1993). This is the result of the infinite number of interacting state variables, which can give rise to a highly complex dynamical behavior. This behavior often can not be described adequately by a low-dimensional chaotic model. Since coupled map lattices are among the simplest models for spatially extended systems they are often used to study this complex behavior. By definition, they have discrete space and time variables and are defined in terms of an evolution rule for the state variables x_n^i, where i denotes the site and n the time. An often used type of evolution rule (see e.g. the review of coupled map lattices by Kaneko (1992)) is

$$x_{n+1}^i = f\left((1 - \epsilon)x_n^i + \frac{\epsilon}{2}(x_n^{i-1} + x_n^{i+1})\right). \tag{9.1}$$

Here we consider spatially extended (infinite) systems by first examining Equation (9.1) for finite systems and taking the limit where the system size becomes infinite. It is a matter of convention whether we call the infinite systems deterministic. In any case, the reconstruction theorem of Takens applies to the finite systems. In the infinite system size limit, for certain functions f, a spectrum of Lyapunov exponents is shown to exist. We then express the information entropy density in terms of the spectrum of Lyapunov exponents. If the information entropy density in the infinite system size limit is positive we speak of spatio-temporal chaos.

A first motivation of this study has been the question whether it is, in principle, possible to distinguish the effect of spatial extension from noise. A second motivation is the following. In the recent literature, conjectures can be found concerning relations between the dimension density and entropy density on the one hand, and spatial structure or structure within a time series on the other hand (Tsimring, 1993; Bauer *et al.*, 1993). It is instructive to test these conjectures by confronting them with exact results.

2. A Solvable Coupled Map Lattice

In this section, a coupled map lattice model is introduced for which a number of analytical results are derived. Consider the model obtained by using the map

$$f(y) = \alpha y \bmod 1 \qquad (9.2)$$

with Equation (9.1) so that

$$x_{n+1}^i = \alpha((1 - \epsilon)x_n^i + \frac{\epsilon}{2}(x_n^{i-1} + x_n^{i+1})) \bmod 1. \qquad (9.3)$$

For a large set of parameters, spatio-temporal chaotic behavior is found in this model. The space-time behavior with the choices of 3 and $\frac{2}{3}$ for α and ϵ respectively is illustrated in Figure 9.1. The model behaves in a complex manner both in time and in space and the spatio-temporal structure is hard to observe by eye. We will focus on large systems; the behavior of the model with a small number of coupled elements and a value of α of 2 was studied by Keller *et al.* (1992).

2.1. Lyapunov Spectra

The Kaplan-Yorke dimension D_{KY} and the information entropy K_1 of a system can be expressed in terms of the ordered Lyapunov exponents, $\lambda_1 \geq \lambda_2 \geq \ldots \geq \lambda_N$, by using the Kaplan-Yorke and Pesin relations (Grassberger *et al.*, 1991; Kaplan and Yorke, 1979). It is conjectured (Farmer *et al.*, 1983) that the Kaplan-Yorke dimension D_{KY} and the information dimension D_1 are identical under general conditions.

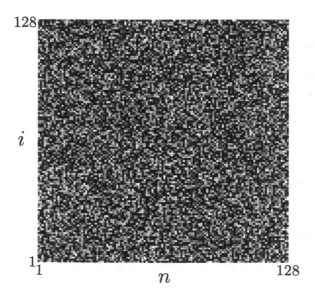

Figure 9.1: *Typical space-time plot of the model given in Equation (9.3) with parameters values $\alpha = 3$ and $\epsilon = \frac{2}{3}$. The grey scales at site i and time n correspond to the values x_i^n. The spatio-temporal structure is hard to observe in the space-time plot.*

The relation between the information dimension D_1 and the Lyapunov exponents then is the following. If there are no positive Lyapunov exponents, the information dimension density is zero. If there are positive Lyapunov exponents, but the sum of all Lyapunov exponents is smaller than zero, then the information dimension D_1 is given by

$$D_1 = j + \frac{\sum_{i=1}^{j} \lambda_i}{|\lambda_{j+1}|} \tag{9.4}$$

where j is the largest integer with $\sum_{i=1}^{j} \lambda_i > 0$. If there are positive Lyapunov exponents, and the sum of all Lyapunov exponents is nonnegative, the information dimension is equal to N, the dimension of the dynamical system. The information entropy K_1 is given by

$$K_1 = \sum_{i:\lambda_i > 0} \lambda_i. \tag{9.5}$$

Anticipating on taking the limit $N \to \infty$, it is convenient to define the information dimension density d_1 as

$$d_1 = \lim_{N \to \infty} \frac{D_1}{N}. \tag{9.6}$$

and the information entropy density k_1 as

$$k_1 = \lim_{N \to \infty} \frac{K_1}{N}. \tag{9.7}$$

Since there is no spatial scale involved but only lattice sites it might be more appropriate to use the terms information dimension per site and information entropy per site but confusion is avoided by defining the unit spatial distance to equal the distance between two nearest neighbor lattice sites.

The piecewise linearity of Equation (9.3) allows an exact calculation of the spectrum of Lyapunov exponents of our model in the case of N sites with periodic boundary conditions. The linear evolution equation of an infinitesimal perturbation ξ_n^i of x_n^i is

$$\xi_{n+1}^i = \alpha \left((1 - \epsilon)\xi_n^i + \frac{\epsilon}{2}(\xi_n^{i-1} + \xi_n^{i+1}) \right), \tag{9.8}$$

(cf. Nicolis *et al.* (1992)), who consider a related coupled map lattice). Upon introduction of the Fourier transform $\hat{\xi}_n^k$ of the perturbation,

$$\hat{\xi}_n^k = \frac{1}{N} \sum_{j=0}^{N-1} \exp(2\pi \imath j k/N)\xi_n^j, \tag{9.9}$$

the Fourier modes can be seen to be eigenvectors of Equation (9.8):

$$\hat{\xi}_{n+1}^k = \alpha \left(1 + \epsilon(\cos(2\pi k/N) - 1) \right) \hat{\xi}_n^k. \tag{9.10}$$

The eigenvalues Γ_k thus are given by

$$\Gamma_k = \alpha \left(1 + \epsilon(\cos(2\pi k/N) - 1) \right). \tag{9.11}$$

The eigenvalues corresponding to the spectrum Γ_k can be visualized as the real parts of N values drawn on a circle in the complex plane with center $\alpha(1 - \epsilon)$ and radius $\alpha\epsilon$. In Figure 9.2 these values are shown for $N = 25$ and the model parameters $\alpha = 3$ and $\epsilon = \frac{2}{3}$.

The Lyapunov exponents associated with Γ_k are $\ln(|\Gamma_k|)$. Note that the Lyapunov exponents associated with the eigenvalues given in Equation (9.11)

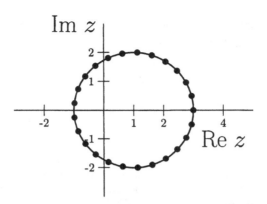

Figure 9.2: *Visualization of the spectrum of eigenvalues for $\alpha = 3$ and $\epsilon = \frac{2}{3}$. For systems of finite size N, the N eigenvalues Γ_k (see Equation (9.11)) correspond to the real parts of N points on a circle in the complex z-plane with center $(\alpha(1 - \epsilon), 0)$ and radius $\alpha\epsilon$. The dots denote the values for a system consisting of $N = 25$ sites. If N approaches infinity, the eigenvalues $\gamma(s)$ (see Equation (9.12)) uniformly fill the circle indicated by the solid line.*

are not necessarily arranged in a non-increasing order for increasing k. In order to obtain the spectrum of ordered Lyapunov exponents we will first consider the asymptotic spectrum of eigenvalues. From this we then construct the spectrum of ordered Lyapunov exponents. In the limit $N \to \infty$, the spectrum of eigenvalues $\gamma(s) = \Gamma_{\{sN\}}$ becomes

$$\gamma(s) = \alpha \left(1 + \epsilon(\cos(2\pi s) - 1)\right) \qquad \text{for } s \in [0, 1). \qquad (9.12)$$

In the limit $N \to \infty$ the eigenvalues corresponding to the spectrum $\gamma(s)$ can again be visualized as the real parts of the values of a circle in the complex plane with center $\alpha(1 - \epsilon)$ and radius $\alpha\epsilon$. In the limit $N \to \infty$, the values uniformly fill the circle (see Figure 9.2). Due to the symmetry of the spectrum of eigenvalues with respect to mirror reflection in the real axis, which amounts to the transformation $s \mapsto (1-s)$, we can conveniently work with the spectrum of eigenvalues $\gamma'(s)$ given by

$$\gamma'(s) = \alpha \left(1 + \epsilon(\cos(\pi s) - 1)\right) \qquad \text{for } s \in [0, 1). \qquad (9.13)$$

For the spectrum of Lyapunov exponents only the density of the values assumed by $\gamma(s)$ and $\gamma'(s)$ are relevant and these are identical.

Furthermore, the spectrum $\gamma'(s)$ of eigenvalues given in Equation (9.13) is invariant with respect to the simultaneous transformation of α, ϵ and s given by

$$(\alpha, \epsilon, s) \mapsto (-\alpha(2\epsilon - 1), \epsilon/(2\epsilon - 1), 1 - s). \tag{9.14}$$

For values of $\epsilon \notin [0, 1]$, we have $\epsilon/(2\epsilon - 1) \in [0, 1]$, so that without loss of generality we may assume $\epsilon \in [0, 1]$. The case $\epsilon = 0$ is best treated separately because the density of eigenvalues is not smooth in this case: all Lyapunov exponents are equal to $\ln(|\alpha|)$. Note that the transformation of Equation (9.14) is an involution, that is, two consecutive applications of the transformation constitute the identity.

The spectrum of Lyapunov exponents is also invariant with respect to the involution

$$\alpha \mapsto -\alpha, \tag{9.15}$$

which reverses all eigenvalues in sign but leaves the spectrum of Lyapunov exponents unchanged. Therefore, for the calculation of the spectrum of Lyapunov exponents we may assume without loss of generality that $\alpha > 0$. The case $\alpha = 0$ is easily treated separately: after one iteration all state variables x_n^i are zero and the system has reached a stable fixed point. In summary, without loss of generality we consider the case $\alpha > 0$, $\epsilon \in (0, 1]$.

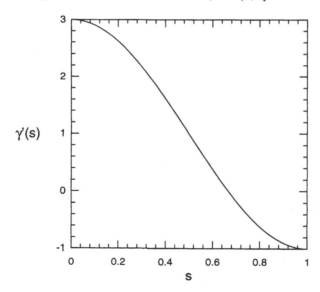

Figure 9.3: *The spectrum $\gamma'(s)$ of eigenvalues for $\alpha = 3$ and $\epsilon = \frac{2}{3}$.*

The next step consists of the construction of $\bar{\gamma}(s)$, the spectrum of ordered absolute values of $\gamma'(s)$. The problem is that, depending on parameter values, identical Lyapunov exponents are associated with values of $\gamma'(s)$ that are identical in absolute value, but differ in sign. The situation is sketched in Figure 9.3. In all cases, the function $\gamma'(s)$ is non-increasing on $[0, 1)$. If $\gamma'(s)$ does not change sign, as is the case if $0 < \epsilon \leq \frac{1}{2}$, the problem described above is not present and we have $\bar{\gamma}(s) = \gamma'(s)$. On the other hand, if $\frac{1}{2} < \epsilon \leq 1$, $\gamma'(s)$ does change sign and $\bar{\gamma}(s)$, for certain values of s, must take into account contributions from the positive as well as the negative values of $\gamma'(s)$.

In Appendix C.1, an expression is derived for the inverse function $\bar{\gamma}^{-1}(s)$.

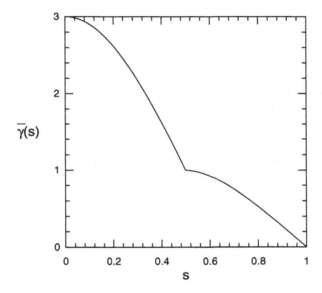

Figure 9.4: *The spectrum $\bar{\gamma}(s)$ of eigenvalues (in absolute value) for $\alpha = 3$ and $\epsilon = \frac{2}{3}$.*

We visualize $\bar{\gamma}(s)$ by using the expression for the inverse function $\bar{\gamma}^{-1}(s)$ given in Equation (C.2). In Figure 9.4, an example of the function $\bar{\gamma}(s)$ is shown for the parameter values $\alpha = 3$ and $\epsilon = \frac{2}{3}$. The corresponding spectrum of Lyapunov exponents, $\lambda(s)$, is given by the relation

$$\lambda(s) = \ln(\bar{\gamma}(s)). \tag{9.16}$$

Figure 9.5 shows the spectrum of Lyapunov exponents for the parameter values $\alpha = 3$ and $\epsilon = \frac{2}{3}$.

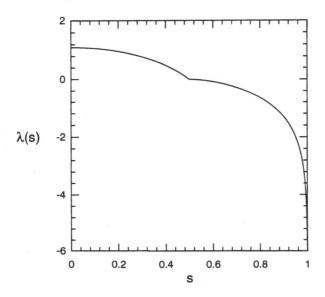

Figure 9.5: *The spectrum* $\lambda(s) = \ln(\bar{\gamma}(s))$ *of Lyapunov exponents for* $\alpha = 3$ *and* $\epsilon = \frac{2}{3}$.

2.2. Dimension Density

In the derivation of the information dimension density of the model we first consider the two exceptional cases, $\alpha = 0$ and $\epsilon = 0$. For $\alpha = 0$, after a single time step the evolution is mapped to a stable fixed point of information dimension 0, so that the information dimension density d_1 is equal to zero. For $\epsilon = 0$, the situation is a little more complicated. All eigenvalues are equal to $\ln(|\alpha|)$. If $|\alpha|$ is not equal to one there is no problem. For $|\alpha| < 1$, all Lyapunov exponents are negative and $d_1 = 0$. For $|\alpha| > 1$, all Lyapunov exponents are positive and $d_1 = 1$. In the case $|\alpha| = 1$, all Lyapunov exponents are 0, and without examining the model behavior care must be taken in the application of the Kaplan-Yorke conjecture. Inspection of the structure of the model shows that it either is in a marginally stable fixed point or exhibits marginally stable period-two behavior. It appears natural to associate an information dimension d_1 of 0 with this situation.

If the information dimension density d_1 is larger than 0 but smaller than 1, it is given by the unique value of h for which

$$\int_0^h \lambda(s)\,\mathrm{d}s = 0, \qquad 0 < h < 1 \tag{9.17}$$

is satisfied. If Equation (9.17) has no solution, then either $d_1 = 0$ (if there are no positive Lyapunov exponents) or $d_1 = 1$. The integration of $\lambda(s)$ in Equation (9.17) is difficult because we do not have an explicit expression for $\lambda(s)$. We will, however, be able to determine the parameter values for which the information dimension density has a value 0, between 0 and 1, and 1.

For various choices of model parameters, $\bar{\gamma}(s)$ tends to zero in the limit where $s \to 1$. The spectrum $\lambda(s)$ then diverges to $-\infty$ for $s \to 1$. The spectrum shown in Figure 9.5 is a typical example showing such a divergence. If this divergence is strong enough, then it could be used to prove that the information dimension density is smaller than 1. Examination of the nature of the divergence, however, shows that it generally is not strong enough. The following argument demonstrates this.

If ϵ is larger than $1/2$, $\gamma'(s)$ has a zero-crossing with a nonzero derivative, as follows from Equation (9.13). Then the spectrum of ordered absolute eigenvalues, $\bar{\gamma}(s)$, approaches 0 algebraically as $s \to 1$, so that the spectrum $\lambda(s)$ diverges logarithmically to $-\infty$ near $s = 1$. The integral in Equation (9.17), however, remains finite in the limit $h \to 1$, because a logarithmically diverging function has a finite integral. Therefore, the logarithmic divergence to $-\infty$ of the spectrum $\lambda(s)$ near $s = 1$ in general is not sufficient to guarantee an information dimension density d_1 smaller than 1.

We can, however, determine whether the information dimension density d_1 is 1 or smaller than 1 by the following observation. The information dimension density d_1 is larger than 0 and smaller than 1 if and only if Equation (9.17) has a solution for $0 < h < 1$. Since $\lambda(s)$ is non-increasing, there can be at most one zero-crossing of the integral in Equation (9.17) in the domain $0 < h < 1$. Thus there is no solution of Equation (9.17) if $I > 0$, where

$$I = \int_0^1 \lambda(s)\,\mathrm{d}s, \tag{9.18}$$

and vice versa, if $d_1 = 1$, there is no solution of Equation (9.17).

The integral I in Equation (9.18) takes into account the absolute values of all Lyapunov exponents with their relative densities in the spectrum. We could have alternatively defined it as the average of the absolute value of the

Lyapunov exponents in the limit of an infinitely large system, viz.

$$I = \lim_{N \to \infty} \frac{1}{N} \sum_{i=1}^{N} \lambda_i, \qquad (9.19)$$

where λ_i is the ith Lyapunov exponent. Because this is independent of the order of summation the integral I in Equation (9.19) can be obtained from the (unordered) spectrum $\gamma'(s)$. We then find

$$\begin{aligned} I &= \int_0^1 \ln(|\,\gamma'(s)\,|) \, ds \\ &= \int_0^1 \ln(|\,\alpha(1-\epsilon) + \epsilon\alpha)\cos(\pi s)\,|) \, ds. \end{aligned} \qquad (9.20)$$

Note that this identity holds for all values of ϵ and α. The above restrictions on α and ϵ were only made for computational convenience, exploiting the invariance of the spectrum of Lyapunov exponents with respect to the transformations given in Equations (9.14) and (9.15). Using Equation (9.20), it can easily be checked that the integral I is also invariant with respect to the transformations given in Equations (9.14) and (9.15).

Upon introducing the variables

$$\begin{aligned} p &= \alpha(1-\epsilon) \\ q &= \alpha\epsilon, \end{aligned} \qquad (9.21)$$

we obtain

$$I = \int_0^1 \ln(|\,p + q\cos(\pi s)\,|) \, ds. \qquad (9.22)$$

The invariance of I with respect to the transformations in Equations (9.14) and (9.15) is represented by invariance with respect to $q \mapsto -q$ and $p \mapsto -p$ respectively. We find

$$I = \begin{cases} \ln\left(|p| + \sqrt{(p^2 - q^2)}\right) - \ln 2 & \text{for } |p| \geq |q| \\ \ln(|q|) - \ln 2 & \text{for } |p| < |q|, \end{cases} \qquad (9.23)$$

a derivation of which is given in Appendix C.2.

The curve on which I is equal to 0 is given by $|q| = 2$ where $|p| < |q|$, and by $|p| = 1 + q^2/4$ where $(|p| \geq |q|, |p| \leq 2)$. The curve encloses a region where $I < 0$ and outside of which we have $I > 0$ (see Figure 9.6). Therefore, in the

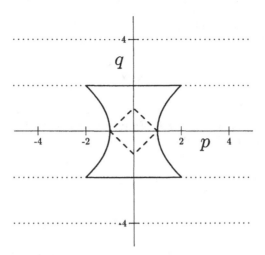

Figure 9.6: *Phase diagram in the p-q plane. The information dimension density is smaller than 1 in the region enclosed by the solid curves and equal to 1 outside this region. The region enclosed by the dashed lines corresponds to parameter values where there are no positive Lyapunov exponents and the information dimension density and the information entropy density are 0. The dotted lines correspond to the parameter values given by $q = 2k$ where k is integer. For these parameter values, the properties of the time series and the spatial series are derived in section 4. Note that the points on the line $q = -p$ are inaccessible by the model. Approaching this line, ϵ tends to $\pm\infty$ as follows from Equation (9.26).*

enclosed region, the information dimension density d_1 is smaller than 1, and outside this region the information dimension density is 1. For all points on the closed curve, we have $I = 0$, which appears to imply $d_1 = 1$. There are however two point-like holes in the curve where it crosses the line $q = 0$. The curve was found under the assumption that ϵ and α are both unequal to zero. The two exceptional points correspond to $\epsilon = 0$ together with $\alpha = \pm 1$, and have $d_1 = 0$ as argued in the beginning of this section.

2.3. Entropy Density

The information entropy density k_1 is given by

$$k_1 \;=\; \int_0^1 \lambda(s)\,\Theta(\lambda(s))\,\mathrm{d}s$$

$$= \int_0^1 \ln(\bar{\gamma}(s)) \, \Theta(\ln(\bar{\gamma}(s))) \, ds \qquad (9.24)$$

$$= \int_0^1 \ln(|\,\gamma'(s)\,|) \, \Theta(\ln(|\,\gamma'(s)\,|)) \, ds$$

where $\Theta(\cdot)$ denotes the Heaviside function

$$\Theta(s) = \begin{cases} 0 & \text{for } s < 0 \\ 1 & \text{for } s \geq 0. \end{cases} \qquad (9.25)$$

The information entropy density is larger than 0 if and only if the spectrum of eigenvalues given in Equation (9.13) has one or more ranges in the interval $0 \leq s \leq 1$ where the eigenvalues are larger than 1 in absolute value. The eigenvalues given in Equation (9.13) range from $\alpha(1 - 2\epsilon) = p - q$ (for $s = 1$) to $\alpha = p + q$ (for $s = 0$). At least one of these limits is larger than 1 in absolute value if either $|p + q| > 1$ or $|p - q| > 1$. This is the case if and only if $|p| + |q| > 1$. The information entropy density, k_1, thus is larger than 0 if and only if $|p| + |q| > 1$. Furthermore, for all finite values of p and q, the Lyapunov exponents are finite which implies that the entropy density k_1 is finite. Figure 9.6 summarizes the different phases in the (p, q) plane.

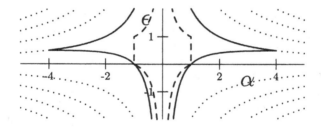

Figure 9.7: *Phase diagram in the α-ϵ plane. The phase diagram is related to the phase diagram in Figure 9.6 through the transformation between the α-ϵ plane and the p-q plane given in Equation (9.21) and the inverse transformation, given in Equation (9.26). For the meaning of the lines and enclosed regions, see Figure 9.6. In the p-q plane, both the solid and the dashed curve are closed, see Figure 9.6. This is no longer the case here and depending on how the line $q = -p$ is approached on either the solid or the dashed curves in the p-q plane, the corresponding points in the α-ϵ plane diverge to $(0, -\infty)$ or $(0, \infty)$.*

Upon inverting Equation (9.21) it is found that the α and ϵ can be expressed in terms of p and q as

$$\begin{aligned} \alpha &= p+q \\ \epsilon &= \frac{q}{p+q}. \end{aligned} \qquad (9.26)$$

The phase diagram in the α-ϵ plane is shown in Figure 9.7.

By definition, the system in the infinite size limit is called spatio-temporally chaotic if the information entropy density is positive. As shown in Figure 9.7, a positive information entropy density is found in a part of parameter space with positive Lebesgue measure (or volume). Now that we have identified the regions in parameter space for which the model exhibits spatio-temporal chaos, we will study the spatial and temporal structure of the model. To this end, in the next section an infinite system is examined which can be considered as the infinite size limit of the systems considered in the calculation of the spectrum of Lyapunov exponents. We note that the derived properties of the Lyapunov spectra are independent of the initial conditions since the evolution of an infinitesimal perturbation of the state, described by Equation (9.8), does not depend on the state itself.

3. Example Evolutions

Now that we have the phase diagram of the model, we can carry out a somewhat more systematic exploration of its behavior. We can choose the parameters to be in any of the three regions found: the non-chaotic region, for which the model evolves to a stable steady state, or spatio-temporal chaos with a positive correlation dimension density. While for $0 < d_1 < 1$ we will speak of spatio-temporal chaos, we will refer to the case with $d_1 = 1$ as fully developed spatio-temporal chaos. In Figure 9.1 above, we showed a space-time pattern for the parameters $\alpha = 3$ and $\epsilon = \frac{2}{3}$, an example of fully developed spatio-temporal chaos. The spatio-temporal structure was not discernible by the naked eye, as a result of the high dimension density. Before returning to the fully developed spatio-temporal chaos in the next section, let us briefly explore the parameter regions in which we have $0 < d_1 < 1$. For small values of d_1 we can expect to see the spatio-temporal structure in the corresponding space-time patterns much easier than for the case $d_1 = 1$. The parameter values for which d_1 is small are expected to lie just outside the dashed areas in the phase diagrams shown in Figures 9.6 and 9.7.

Figure 9.8 shows space-time plots for $\epsilon = 0.5$ for several values of α. Given the value of ϵ the spectrum of Lyapunov exponents depends only on the absolute value of α. The four plots in Figure 9.8 correspond to two pairs of cases

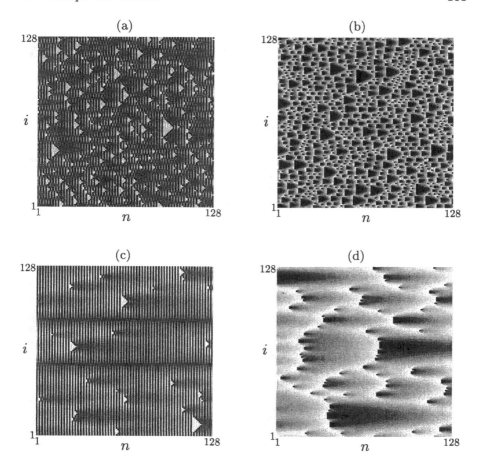

Figure 9.8: *Space-time plots for* $\epsilon = \frac{1}{2}$, *with (a)* $\alpha = -1.2$, *(b)* $\alpha = 1.2$, *(c)* $\alpha = -1.01$ *and (d)* $\alpha = 1.01$. *Initial conditions were chosen according to the uniform distribution on* $[0, 1)$ *and transients were discarded. Note that cases (a) and (b) have the same spectra of Lyapunov exponents. The same holds for cases (c) and (d).*

with the same absolute value of α and opposite signs. For such a pair the spectra of Lyapunov exponents, and hence the correlation dimension densities and correlation entropy densities are the same. The spatio-temporal patterns bear some resemblance to the space-time patterns found in the cellular automaton models described by Wolfram (1984). Although each of the two neighboring patterns have identical spectra of Lyapunov exponents, the patterns do not re-

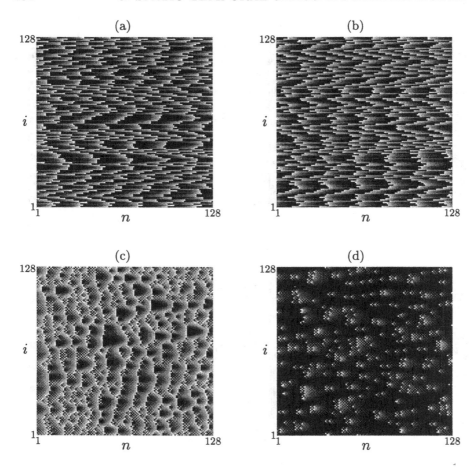

Figure 9.9: *Space-time plots for (a) $p = 1$, $q = 1/10$ ($\alpha = 11/10$, $\epsilon = 1/11$), (b) $p = 1$, $q = -1/10$ ($\alpha = 9/10$, $\epsilon = -1/9$), (c) $p = 1/10$, $q = 1$ ($\alpha = 11/10$, $\epsilon = 10/11$) and (d) $p = -1/10$, $q = 1$ ($\alpha = 9/10$, $\epsilon = 10/9$). Initial conditions were chosen according to the uniform distribution on $[0, 1)$ and transients were discarded. As for Figure 9.8, cases (a) and (b) have the same spectra of Lyapunov exponents. The same holds for cases (c) and (d).*

semble each other much. We could have anticipated this, as equivalence of the spectrum of Lyapunov exponents does not imply conjugation of the dynamics.

Figure 9.9 shows space-time plots for some other values of α and ϵ. The four plots in Figure 9.9 also correspond to two pairs with the same spectra of

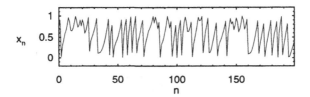

Figure 9.10: *Segment of a time series measured at a single site for* $\alpha = 1.2$ *and* $\epsilon = 0.5$.

Lyapunov exponents, and hence with identical correlation dimension density and correlation entropy density. The space-time patterns for the parameter values $p = 1$, $q = 1/10$ ($\alpha = 11/10$, $\epsilon = 1/11$), and $p = 1$, $q = -1/10$ ($\alpha = 9/10$, $\epsilon = -1/9$), appear to be similar, apart from a reversion of the time direction. The model equations are

$$x_{n+1}^i = x_n^i + \delta(x_n^{i-1} + x_n^{i+1}) \bmod 1 \tag{9.27}$$

with $\delta = 1/20$ for the case $p = 1$, $q = 1/10$, and $\delta = -1/18$ for the case $p = 1$, $q = -1/10$. Now let us consider the evolution for a fixed but small value of δ described by Equation (9.27). Upon time reversal, i.e. interchanging $n + 1$ and n, we get

$$x_n^i = x_{n+1}^i + \delta(x_{n+1}^{i-1} + x_{n+1}^{i+1}) \bmod 1, \tag{9.28}$$

or equivalently

$$x_{n+1}^i = x_n^i - \delta(x_{n+1}^{i-1} + x_{n+1}^{i+1}) \bmod 1. \tag{9.29}$$

Note that the change of each state per time step is of the order δ, so that we may write

$$x_{n+1}^i = x_n^i - \delta(x_n^{i-1} + x_n^{i+1}) + O(\delta^2). \tag{9.30}$$

Time reversal thus is equivalent, up to order δ, to changing the sign of δ. This explains why the models with $\delta = 1/20$ and $\delta = -1/18$ which are small and nearly the same in absolute value, give space-time patterns that appear to be identical apart from time reversal.

A segment of length 200 from a time series measured at a single site for $\alpha = 1.2$ and $\epsilon = 0.5$ is shown in Figure 9.10. The phase portrait of a longer time series with $N = 4000$ generated for these parameter values is shown in Figure 9.11. Note the clear indication of dependence between consecutive measurements. The non-uniqueness of x_{n+1}^i given x_n^i observed in the phase plot shows that the time series can not be described well by a first order deterministic map.

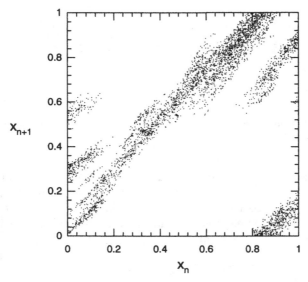

Figure 9.11: *Delay plot for a time series measured at a single site for $\alpha = 1.2$ and $\epsilon = 0.5$. A time series with length $N = 4000$ is used.*

3.1. Fitting the Noisy Attractor Model

As remarked in Chapter 6, a possible explanation for the fact that the noisy attractor model does not fit the Gaussian kernel correlation integrals of the fluidized bed pressure time series very well is the spatio-temporal nature of the fluidized bed. Due to spatio-temporal interactions, locally measured time series in such a system are unlikely to behave as low-dimensional time series with observational noise. To examine what happens when we fit the noise model to a typical time series from a spatio-temporally chaotic system, we examine a time series from our model for $\alpha = 1.2$ and $\epsilon = 0.5$, a segment and a phase portrait of which are shown in Figures 9.10 and 9.11 respectively. The corresponding estimates of σ, D_2 and K_2 are shown in Figure 9.12. These resemble the estimates (see Figure 6.8) for the fluidized bed time series in that the estimated correlation dimension increases with the embedding dimension, while the estimated correlation entropy saturates for larger values of the embedding dimension.

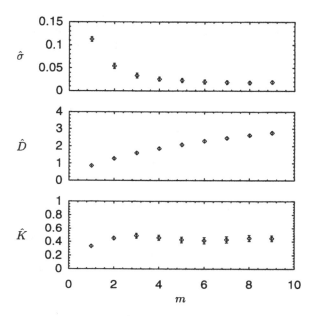

Figure 9.12: *Estimated values $\widehat{\sigma}$, $\widehat{D}_2$ and $\widehat{K}_2$ as a function of m for the time series from the spatio-temporal coupled map lattice model. The bars denote the estimated 95% confidence interval (2 standard errors).*

4. Spatial and Temporal Structure

In this section, we will first investigate the properties of spatial series $\{S_i\}$ of the model and then those of time series $\{T_n\}$ measured at a single site. The measurement function is taken to be the identity, so that

$$S_i = x_n^i, \qquad \text{for } n \text{ fixed.} \tag{9.31}$$

For certain parameter values α and ϵ, it can be shown that the spatial distribution where the x_i^n are independent, identically distributed (i.i.d.) random variables, each x_n^i being drawn independently from the uniform distribution on $[0, 1)$, is invariant. This is the case when we restrict ourselves to parameters α and ϵ for which

$$\alpha\epsilon = 2k, \tag{9.32}$$

where $k \in \mathbf{Z}, k \neq 0$. In Figures 9.6 and 9.7 these parameter values are indicated by dotted lines. The system size is taken to be infinite and an infinitely large number of sites is extending from each site to plus and minus infinity.

4.1. Spatial Series

Starting at time $n = 0$ with an i.i.d. uniform field as described above, it can be easily shown that the field at time $n = 1$ is also i.i.d. The evolution rule is

$$x_1^i = ((\alpha - 2k)x_0^i + kx_0^{i-1} + kx_0^{i+1}) \bmod 1, \qquad (9.33)$$

which may be written as

$$x_1^i = (F(x_0^i, x_0^{i-1}) + kx_0^{i+1}) \bmod 1. \qquad (9.34)$$

If the modulo operation is regarded as a reminder that all addition operations within the brackets are performed on the one-torus, Equation (9.34) amounts to an addition of the random variable kx_0^{i+1} to some point F on the torus. This operation yields a uniform random variable in the range $[0, 1)$ regardless of the value of F. From Equation (9.34) it follows that x_1^i is independent of all values x_1^j for all $j < i$. The mirror reflection symmetry of the system implies that one could alternatively write

$$x_1^i = (F(x_0^i, x_0^{i+1}) + kx_0^{i-1}) \bmod 1, \qquad (9.35)$$

from which it follows that x_1^i is independent of all values x_1^j for all $j > i$.

Induction with respect to the time variable n shows that an initially i.i.d. spatial field with a uniform distribution remains i.i.d. with a uniform distribution in the course of time. Although this i.i.d. spatial distribution is invariant, the distribution is not necessarily physical in the sense that it is stable under the dynamics with respect to perturbations. However, numerical simulations suggest that there is convergence towards an i.i.d. distribution. For example Figure 9.13 shows the result of a numerical experiment in which the initial spatial pattern at $n = 0$ has an i.i.d. uniform distribution on $[0, 0.0001)$. One can observe a quick convergence towards a complicated spatial field. This numerical experiment suggests that the invariant i.i.d. uniform spatial field on $[0, 1)$ corresponds to the natural measure and we will take the initial conditions accordingly.

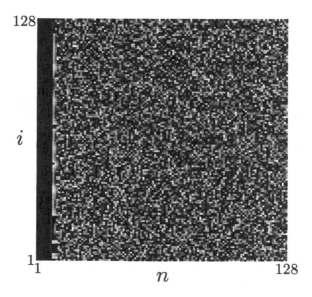

Figure 9.13: *Transient behavior of the model (9.3) with parameters values $\alpha = 3$ and $\epsilon = \frac{2}{3}$. The grey scales at site i and time n correspond to the values x_i^n. The initial values were chosen i.i.d. according to the uniform distribution on $[0, 0.0001)$. After several iterations the pattern becomes highly complicated, cf. Figure 9.1 where we choose i.i.d. uniformly distributed random initial conditions in the range $[0, 1)$.*

4.2. Time Series

We next investigate the properties of the time series $\{T_n\}$ produced by model (9.3) according to

$$T_n = x_n^i, \qquad \text{for } i \text{ fixed.} \tag{9.36}$$

Without loss of generality we assume we are measuring at site $i = 0$.

Consider a measurement at time n. The measurements made before time n depend on the initial conditions of the sites in the backwards 'light cone' (Mayer-Kress and Kaneko, 1989) of site 0. The sites involved are $-n + 1, -n + 2, \ldots, n - 1$. At time n, the initial conditions of site $-n$ and site $+n$ become involved in the measurement at site i. Using Equation (9.3) it can be easily shown that the x_0^{-n} and x_0^n enter the measurement T_n in the following way:

$$T_n = \left(G_n(x_0^{-n+1}, x_0^{-n+2}, \ldots, x_0^{n-1}) + (k^n)x_0^{-n} + (k^n)x_0^n\right) \bmod 1, \tag{9.37}$$

where G_n is some function of its arguments. Here we have used the general relation

$$\left(\sum_{i=1}^{l}(a_i \bmod 1)\right) \bmod 1 = \left(\sum_{i=1}^{l} a_i\right) \bmod 1. \qquad (9.38)$$

The terms x_0^{-n} and x_0^n are independent of G_n. and uniformly distributed in the range $[0,1)$. Equation (9.37) represents the addition of the random variables $(k^n)x_0^{-n}$ and $(k^n)x_0^n$ to some point G_n on the torus. The result is an i.i.d. random variable in the range $[0,1)$ regardless of the value of G_n. This shows that each measurement yields a new random variable T_n that is uniformly distributed in $[0,1)$ and independent of all previous measurements made at the same site. Clearly, the time series has all the properties of an i.i.d. time series with a uniform probability distribution in $[0,1)$.

From the above argument it follows that, as far as measuring at single sites is concerned, there is no difference between measuring from an infinite system and measuring from a finite system with random boundary conditions with a uniform distribution in $[0,1)$. Consider a bounded part of the array of sites, say with left and right boundary sites l and r respectively. At each time n, two additional initial values, x_0^{l-n} and x_0^{r+n}, are influencing the left and right boundary sites. The states x_n^l and x_n^r will be independent of the history of the bounded part and as far as the sites in the bounded part are concerned this state will be indistinguishable from a random state with uniform distribution in $[0,1)$.

For the parameter values resulting in i.i.d. time series and i.i.d. spatial series, the information dimension density was found to be 1 in the previous section. However, when considering these parameter values, the information entropy density may still depend on the parameter values. We have performed a number of numerical experiments, in which Equation (9.13) was used to determine the spectrum of Lyapunov exponents. Figure 9.14 shows the information entropy density k_1 calculated from this spectrum for various values of the parameter α. In each calculation ϵ was chosen so that $\alpha\epsilon = 2$, cf. Equation (9.32), to ensure that in each case the time series and spatial series had identical i.i.d. properties. A number N of 2000 sites was used, a value for which a good convergence of k_1 is obtained (calculations with larger values of N suggest that the relative errors of the estimated values of k_1 are of the order of 10^{-4}). It is clear from Figure 9.14 that k_1 depends on α, although for each value of α the model generates indistinguishable time series and spatial series. It follows that different information entropy densities do not necessary result in time series or spatial series with different properties. The information

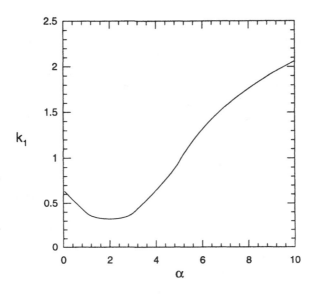

Figure 9.14: *The information entropy density k_1 as a function of α, for $\alpha\epsilon = q = 2$. For all values of α, the model generates indistinguishable time series and spatial series.*

entropy density of the model therefore in general can not be determined from a single scalar time or spatial series only.

5. Summary and Discussion

In this chapter we have examined a coupled map lattice model as a function of the model parameters. The piece-wise linearity of the model allows the exact calculation of the spectrum of Lyapunov exponents and the construction of a phase diagram showing the regions in parameter space where the information dimension density is 0, 1, or between 0 and 1. Also, the regions in parameter space were found where the information entropy density is zero and where it is positive. For parameter values in the latter regions the model exhibits spatio-temporal chaos. For some model parameters, the structure of time series measured at single sites could be determined, and was found to be i.i.d. For these model parameters the spatial series are also i.i.d. Therefore, neither the properties of the spatial series nor of the time series indicate that there is spatio-temporal structure in the model. If a time series or a spatial series would be studied without knowledge of the model, one could be easily led into thinking that the model shows no signs of spatio-temporal structure.

In the literature, a number of candidate methods have been suggested for the characterization of spatio-temporal chaos. We will briefly describe these methods and discuss them in the context of our model.

Gaspard and Wang (1993) proposed a characterization method that consists of determining the coarse-grained entropy of a system. By construction, it is suited for the characterization of a large variety of deterministic and stochastic systems. However, their expression for the coarse-grained entropy of coupled map lattices applies to the entire state space of a finite system, and not to scalar time-series measured at a single site. Therefore we are unable to compare this expression with the coarse-grained entropy of a single scalar time series. The coarse-grained entropy can be determined for a single time series, but applied in this way it characterizes the time series rather than the underlying spatio-temporal model.

Bauer *et al.* (1993) proposed a method for estimating the correlation dimension density d_2 from a single spatial series. For the parameter values of our model with an invariant i.i.d. density, there is no fractal structure, so that we have $d_1 = d_2$. When the method of Bauer *et al.* would be applied to an i.i.d. time series, the value $d_2 = 1$ would be found. For the model parameters for which we found that the time series were i.i.d. this is the correct value.

Tsimring (1993) proposed a form of the correlation integral of a time series measured at a single position in which the correlation dimension density d_2 and correlation entropy density k_2 occur as parameters. The method is designed for estimation of both the correlation dimension density and the correlation entropy density from a single observed local time series by examining correlation integrals. For our model, for the parameter values along the lines $\alpha\epsilon = 2k$, the entropy density depends on the model parameters, but the time series are all identical. For these parameter values, therefore, it is not possible to determine k_2 from the correlation integral of a single time series. More generally this implies that a general method for the complete characterization of spatio-temporal chaos based on the analysis of a single time series or a single spatial series is impossible in principle. Identical spatial or temporal series can be produced by a number of models, with different values of the information entropy density.

The observation that time series with a similar structure can be obtained from models with different information entropy densities suggests that both space and time must be taken into account in order not to miss the essential features of the model. One may think of using a multi-site reconstruction, where an m-dimensional shift vector, such as $(x_n^i, x_n^{i+1}, \ldots, x_n^{i+m-1})$, is followed in the course of time (i.e. for consecutive values of n). The predictors of the vector elements, constructed from vectors observed in the past (cf. Pezard *et al.*, 1996) by construction contain space-time information and could be helpful in characterizing spatio-temporal structure. One may also think of generalizing the concept of delay vectors to delay-shift vectors. For example, in the 4-dimensional vectors $(x_n^{i-1}, x_n^i, x_n^{i+1}, x_{n+1}^i)$, the last component is determined by the first three components by the evolution rule. As a result, the dimension of the probability distribution is at most 3. For a discussion on the generalization of the delay vector concept to spatio-temporal matrices we refer to Abarbanel (1993).

The coupled map lattice model presented here shows that a general method for the complete characterization of spatio-temporal chaos can not be based on the analysis of a single time series or a single spatial series only, as a single time series or a single spatial series generally is not sufficient for the characterization of the model. Similar time series and similar spatial series were produced for various model parameters with different values of the information entropy density k_1.

It is fair to say, however, that the system studied is somewhat unphysical in that it has discontinuities in the evolution rule (the modulo function). At present we are not aware of a spatially extended dynamical system with similar properties, without discontinuities in its evolution rule. Furthermore, although our model generates similar time and spatial series for different model parameters which give rise to different space-time behavior, it does so only in a subset of zero Lebesgue measure in parameter space.

The result that the invariant measure is uniform, which we derived for integer $k = \alpha \epsilon / 2$, can be generalized to a lattice of Bernoulli maps coupled by a matrix with integer coefficients (Grigoriev, 1998). In contrast to infinite systems, systems of finite size do possess a nontrivial coherence structure when the linearized dynamics has contracting directions. Upon examining the linear correlation functions of finite systems, it is concluded that the notion of linear correlation is insufficient to reveal the coherence in systems beyond a certain minimum size. Higher order correlation functions or nonlinear correlation measures are necessary to reveal it. It would be reasonable to expect that this nontrivial coherence becomes more difficult to detect if the system size gets larger, as in this limit the spatial patterns become independent, identically distributed, just as in our model.

Although exact results can be found by using solvable models such as those of Grigoriev (1998) and the one described in this chapter, they are solvable for specific parameter values, corresponding to fully developed spatio-temporal chaos ($d_1 = 1$) only, at least in the limit of infinitely large systems. It would be interesting to have exact results for parameter values for which more space-time structure is present (such as for the example evolutions shown with $d_1 <$ 1). For these parameter values the spatio-temporal structure is much more difficult to trace, as the invariant measure will no longer have a simple uniform structure.

Appendix A

Reversibility and Dynamical Systems

In Chapter 4 a definition of reversibility was given for time series. The notion of reversibility of dynamical system is also known. In this appendix we discuss the connections between the two notions of reversibility. Throughout, we will consider only discrete time dynamical systems. We are interested in the question whether there is a relation between reversible dynamical systems and reversible time series. Some preliminary remarks are appropriate here. Firstly, any dynamical system can generate reversible time series, as is clearly demonstrated by the trivial example of a measurement function which is constant over state space. Secondly, it can be readily checked that reversible time series are obtained for all measurement functions when the evolution in state space has been attracted to a fixed point or to a period-2 attractor. Following Roberts and Quispel (1992), we define reversibility of a discrete time dynamical system $(\Omega, F, \mathbf{Z})$ as follows.

Definition A.1 *A discrete time dynamical system with evolution map F is reversible if there exists an involution H, i.e. a map H with the property $H \circ H = I$, such that*

$$H \circ F \circ H^{-1} = F^{-1}. \tag{A.1}$$

Note that a reversible dynamical system is necessary invertible. The solutions of Equation (A.1) with respect to H will be referred to as reversing maps. Given such a solution, H_1 say, the dynamical system is said to be reversible under H_1.

It is not difficult to formulate a number of conditions (concerning the dynamics and the measurement function) under which a reversible dynamical system gives (nontrivial) reversible time series. In fact, this is what we will do below for a larger class of dynamical systems, namely that of *weakly reversible* dynamical systems.

Definition A.2 *A discrete time dynamical system with evolution map F is weakly reversible if there exists an invertible map G, such that*

$$G \circ F \circ G^{-1} = F^{-1}. \tag{A.2}$$

The solutions of Equation (A.2) with respect to G are called weakly reversing maps. Given such a solution, G_1 say, the dynamical system is said to be weakly reversible under G_1. With the latter definition, we are in a position to derive the following result:

Proposition A.1 *If the measurement function h is invariant under at least one weakly reversible map that preserves the invariant measure of the dynamical system, the time series obtained with h is reversible.*

Proof: An m-dimensional delay vector

$$\boldsymbol{X}_n = (h(\boldsymbol{x}_n), h(F\boldsymbol{x}_n), \ldots, h(F^{m-1}\boldsymbol{x}_n)), \tag{A.3}$$

by G-invariance of h can be written as

$$\boldsymbol{X}_n = (h(G\boldsymbol{x}_n), h(GF\boldsymbol{x}_n), \ldots, h(GF^{m-1}\boldsymbol{x}_n)). \tag{A.4}$$

Application of Equation (A.2) gives

$$\boldsymbol{X}_n = (h(G\boldsymbol{x}_n), h(F^{-1}G\boldsymbol{x}_n), \ldots, h(F^{1-m}G\boldsymbol{x}_n)), \tag{A.5}$$

which by G-invariance gives

$$\boldsymbol{X}_n = (h(F^{m-1}\boldsymbol{y}_n), h(F^{m-2}\boldsymbol{y}_n), \ldots, h(F\boldsymbol{y}_n)), \tag{A.6}$$

with $\boldsymbol{y}_n = F^{1-m}G\boldsymbol{x}_n$. By the assumed measure preserving properties of F and G, $\boldsymbol{x}_n$ and $\boldsymbol{y}_n$ have identical measures. It follows that $\boldsymbol{X}_n$ and $P\boldsymbol{X}_n$ have identical probability measures, and hence that the time series is reversible. $\square$

Example: Anosov Map Consider the Anosov diffeomorphism on the 2-torus, given by

$$\left(\begin{array}{c} x_{n+1} \\ y_{n+1} \end{array} \right) = F \left(\begin{array}{c} x_n \\ y_n \end{array} \right) \tag{A.7}$$

where

$$F \left(\begin{array}{c} x_n \\ y_n \end{array} \right) = \left(\begin{array}{cc} 1 & 1 \\ 1 & 0 \end{array} \right) \left(\begin{array}{c} x_n \\ y_n \end{array} \right) \bmod 1. \tag{A.8}$$

The stable invariant measure is the uniform measure on the torus. The inverse F^{-1} is given by

$$F^{-1} \left(\begin{array}{c} x_n \\ y_n \end{array} \right) = \left(\begin{array}{cc} 0 & 1 \\ 1 & -1 \end{array} \right) \left(\begin{array}{c} x_n \\ y_n \end{array} \right) \bmod 1. \tag{A.9}$$

This Anosov map is irreversible because the eigenvalues of the matrices occurring in F and F^{-1} are different. It is, however, weakly reversible, as Equation (A.2) can be solved for G, giving the two solutions

$$G_1 \left(\begin{array}{c} x_n \\ y_n \end{array} \right) = \left(\begin{array}{c} 1 - y_n \\ x_n \end{array} \right), \quad G_2 \left(\begin{array}{c} x_n \\ y_n \end{array} \right) = \left(\begin{array}{c} y_n \\ 1 - x_n \end{array} \right). \tag{A.10}$$

The maps G_1 and G_2 correspond to rotations around $(x, y) = (\frac{1}{2}, \frac{1}{2})$ over $-\frac{1}{2}\pi$ and $\frac{1}{2}\pi$ respectively, and they both preserve the uniform asymptotic distribution.

If the measurement function is symmetrical with respect to either G_1 or G_2, the observed time series will be reversible. Note that symmetry with respect to a map G implies symmetry with respect to the group induced by G. In this example the groups induced by G_1 and G_2 are the same (the group of rotations around $(1/2, 1/2)$ over multiples of $2\pi/4$), so that symmetry of the measurement function with respect to G_1 implies symmetry with respect to G_2 and vice versa.

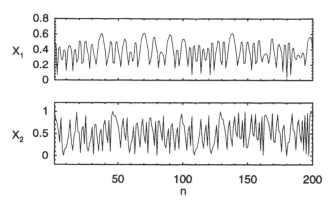

Figure A.1: *Two time series generated with the Anosov diffeomorphism. Time series $X_1(n)$ (upper panel) is generated with the symmetric measurement function h_1 and $X_2(n)$ (lower panel) with the asymmetric measurement function h_2. The two time series are obtained with different random initial conditions.*

It can be readily verified that the measurement function h_1 given by

$$h_1(x_n, y_n) = \sqrt{(1/2 - x_n)^2 + (1/2 - y_n)^2}, \tag{A.11}$$

is symmetric with respect to G_1 and hence G_2. Furthermore, h_2 defined by

$$h_2(x_n, y_n) = x_n, \tag{A.12}$$

is not symmetric with respect to G_1 and G_2.

Two time series $X_1(n)$ and $X_2(n)$ of 200 consecutive discrete times were generated from a numerical implementation of the dynamical system defined by Equation (A.8) using the measurement functions h_1 and h_2 respectively. The time series are shown in Figure A.1. Indeed, as expected, the first time series shows no sign of irreversibility, whereas the second time series appears to be irreversible. We applied our reversibility test to these time series, and for $\{X_1(n)\}$ we found $S = -0.18$ which provides no evidence against the hypothesis of reversibility. The value of the test statistic found for $\{X_2(n)\}$ is $S = 8.31$ which strongly suggests irreversibility.

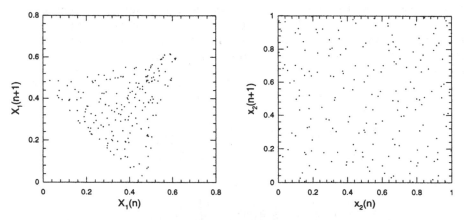

Figure A.2: *Two-dimensional delay vector distributions (phase portraits) of $X_1(n)$ (left panel) and $X_2(n)$ (right panel). Both plots appear to be consistent with symmetry with respect to time reversal (the distributions represented by the dots are symmetric with respect to interchanging the two coordinates). The 3-dimensional delay vector distribution of series $X_2(n)$ was however found to be inconsistent with a reversible time series.*

Figure A.2 shows the corresponding phase portraits of $\{X_1\}$ and $\{X_2\}$. Both phase portraits are symmetrical with respect to reflection in the diagonal line. No signs of irreversibility are found in these phase portraits. In fact, the second time series is a nice example where irreversibility can not be detected in the two-dimensional phase portrait. The irreversibility is detectable only in higher dimensional delay vector distributions. In general, there may be time series for which the irreversibility shows up only above an embedding dimension of k. A convenient term for these time series would be order-k reversible. In analogy with order-k reversibility the time series $X_2(n)$ could be termed an order-2 independent time series, because up to embedding dimension 2 its reconstruction measure is indistinguishable from that of an i.i.d. time series with a uniform marginal distribution on $[0, 1)$. $\square$

Example: Logistic Map For dynamical systems that have no unique inverse evolution operator, like the logistic map, one can also construct measurement functions that produce non-constant reversible time series. For the logistic map an example of such a function is

$$h(x) = \begin{cases} 0 & \text{for } x < \frac{1}{2} \\ 1 & \text{for } x \geq \frac{1}{2}, \end{cases} \tag{A.13}$$

which gives rise to an i.i.d. sequence of zeros and ones, both with equal probability. The measurement function, however, is not smooth. The fact that the time series are i.i.d. is best seen after reversing the direction of time. The logistic map on the interval $[0, 1]$ has no unique inverse, but the time reversed dynamics (Tong and Cheng, 1992; Lawrance and Spencer, 1995) can be described by the stochastic process:

$$x_{n+1} = \begin{cases} \frac{1}{2} + \frac{1}{2}\sqrt{1 - x_n} & \text{with probability } \frac{1}{2} \\ \frac{1}{2} - \frac{1}{2}\sqrt{1 - x_n} & \text{with probability } \frac{1}{2}, \end{cases} \tag{A.14}$$

so that $h(x_n)$ is either 0 or 1, both with probability $\frac{1}{2}$, independently of all other observations. □

In summary, both reversible and irreversible dynamical systems generally give rise to irreversible time series and reversible time series are obtained only with special measurement functions. Fixed point attractors and period-2 attractors are exceptional in that they generate reversible time series for every measurement function.

Appendix B

Variance of the Test Statistic

In this appendix we derive analytic expressions for the unconditional and the conditional variance of the statistic $\widehat{Q}$ which we used in the test for comparing reconstruction measures in Chapter 5. The derivation of the unconditional variance (which is not used in the test) is included for completeness. We will closely follow the methodology of Van Zwet (1984).

Appendix B.1. Unconditional Variance

In order to determine the unconditional variance of $\widehat{Q}$ the random vectors $\{X_i\}_{i=1}^{N_1}$ and $\{Y_i\}_{i=1}^{N_2}$ are assumed to be independent samples from identical probability measures μ_1. We have

$$\widehat{Q} = \frac{1}{\binom{N_1}{2}} \sum_{1 \leq i < j \leq N_1} h(X_i, X_j) + \frac{1}{\binom{N_2}{2}} \sum_{1 \leq i < j \leq N_2} h(Y_i, Y_j)$$
$$- \frac{2}{N_1 N_2} \sum_{i=1}^{N_1} \sum_{j=1}^{N_2} h(X_i, Y_j). \tag{B.1}$$

Denoting expected values by $E(\cdot)$, we define the functions

$$H(x, y) = h(x, y) - E(h(X_1, X_2)), \tag{B.2}$$

and

$$g(x) = E(H(X_1, X_2)\|X_1 = x). \tag{B.3}$$

Now $\widehat{Q}$ can be expressed in terms of

$$\psi(x, y) = H(x, y) - g(x) - g(y), \tag{B.4}$$

by substituting

$$h(x, y) = \psi(x, y) + g(x) + g(y) + E(h(x, y)) \tag{B.5}$$

into Equation (B.1). Since the contributions $g(\cdot)$ and $E(h(\cdot, \cdot))$ sum to zero, we obtain

$$\widehat{Q} = \frac{1}{\binom{N_1}{2}} \sum_{1 \leq i < j \leq N_1} \psi(X_i, X_j) + \frac{1}{\binom{N_2}{2}} \sum_{1 \leq i < j \leq N_2} \psi(Y_i, Y_j)$$
$$- \frac{2}{N_1 N_2} \sum_{i=1}^{N_1} \sum_{j=1}^{N_2} \psi(X_i, Y_j). \tag{B.6}$$

Now, by construction we have

$$E(\psi(\boldsymbol{X}_i, \boldsymbol{X}_j)\|\boldsymbol{X}_i = \boldsymbol{x}) = 0 \qquad \text{for all } \boldsymbol{x}, \tag{B.7}$$

so that the right hand side of Equation (B.6) consists of a sum of terms which all are uncorrelated. Denoting variances by $V(\cdot)$, the variance of $\widehat{Q}$ can be written as

$$V(\widehat{Q}) = (\frac{2}{N_1(N_1 - 1)} + \frac{2}{N_2(N_2 - 1)} + \frac{4}{N_1 N_2})V(\psi(\boldsymbol{X}_1, \boldsymbol{X}_2)). \tag{B.8}$$

We can express $V(\psi(\boldsymbol{X}_1, \boldsymbol{X}_2))$ as

$$V(\psi(\boldsymbol{X}_1, \boldsymbol{X}_2)) = V(h(\boldsymbol{X}_1, \boldsymbol{X}_2)) - 2V(g(\boldsymbol{X}_1)), \tag{B.9}$$

using the definitions given in Equations (B.2)–(B.4).

Appendix B.2. Conditional Variance

To calculate the variance of the statistic $\widehat{Q}$, conditionally on the observed set of reconstruction vectors, we introduce $N = N_1 + N_2$ and define

$$z_i = \begin{cases} \boldsymbol{X}_i & \text{for } 1 \le i \le N_1 \\ \boldsymbol{Y}_{i-N_1} & \text{for } N_1 < i \le N. \end{cases} \tag{B.10}$$

Given the set of vectors $\{z_i\}_{i=1}^N$, we consider random divisions of their indices into two groups of sizes N_1 and N_2. We obtain

$$\begin{aligned}
\widehat{Q} &= \frac{1}{\binom{N_1}{2}} \sum_{\substack{i,j \in D \\ i<j}} \sum h(z_i, z_j) + \frac{1}{\binom{N_2}{2}} \sum_{\substack{i,j \in D^c \\ i<j}} \sum h(z_i, z_j) \\
&\quad - \frac{2}{N_1 N_2} \sum_{i \in D} \sum_{j \in D^c} h(z_i, z_j),
\end{aligned} \tag{B.11}$$

where D is a set of N_1 indices randomly selected without replacement from $\{1, 2, \ldots, N\}$ and D^c is the complementary set of indices. We write Equation (B.11) as

$$\widehat{Q} = \sum_{i<j} \sum h(z_i, z_j) A_{ij}, \tag{B.12}$$

with

$$
\begin{aligned}
A_{ij} \;=\; & \frac{1}{\binom{N_1}{2}} 1_D(i)\,1_D(j) + \frac{1}{\binom{N_2}{2}} 1_{D^c}(i)\,1_{D^c}(j) \\
& - \frac{2}{N_1 N_2} 1_D(i)\,1_{D^c}(j) - \frac{2}{N_1 N_2} 1_{D^c}(i)\,1_D(j),
\end{aligned}
\tag{B.13}
$$

where $1_D(i)$ is the indicator function defined by

$$
1_D(i) = \begin{cases} 1 & \text{if } i \in D \\ 0 & \text{if } i \notin D. \end{cases}
\tag{B.14}
$$

The A_{ij} in Equation (B.12) can be identified as random variables, while the $h(z_i, z_j)$ are constants. The conditional expected value $E_c(\widehat{Q})$ of $\widehat{Q}$ is zero since

$$
E_c(A_{ij}\|1_D(j)) = E_c(A_{ij}\|1_D(i)) = 0
\tag{B.15}
$$

for all i and j. We define the constants

$$
H_{ij} = h(z_i, z_j) - \frac{1}{\binom{N}{2}} \sum_{i'<j'} \sum h(z_{i'}, z_{j'}),
\tag{B.16}
$$

$$
g_i = \frac{1}{N-2} \sum_{\substack{j \\ j \neq i}} H_{ij},
\tag{B.17}
$$

and

$$
\psi_{ij} = H_{ij} - g_i - g_j.
\tag{B.18}
$$

Since A_{ij} is symmetric in i and j and

$$
\sum_{\substack{i \\ i \neq j}} A_{ij} = 0 \qquad \text{for all } j,
\tag{B.19}
$$

the statistic $\widehat{Q}$ remains unchanged by replacing $h(z_i, z_j)$ with ψ_{ij} and we can express $\widehat{Q}$ as

$$
\widehat{Q} \;=\; \sum_{i<j} \sum \psi_{ij} A_{ij}.
\tag{B.20}
$$

The calculation of the conditional variance $Var_c\{\widehat{Q}\}$ which is the conditional expected value of $\widehat{Q}^2$, $E_c(\widehat{Q}^2)$, is straightforward. By counting the terms in $\widehat{Q}^2$

we obtain

$$
\begin{aligned}
Var_c\{\widehat{Q}\} &= \frac{1}{2}N(N-1)E_c(A_{12}^2)\overline{\psi_{ij}^2} \\
&+ N(N-1)(N-2)E_c(A_{12}A_{23})\overline{\psi_{ij}\psi_{jk}} \\
&+ \frac{1}{4}N(N-1)(N-2)(N-3)E_c(A_{12}A_{34})\overline{\psi_{ij}\psi_{kl}},
\end{aligned}
\tag{B.21}
$$

where the bars denote taking averages (with i, j, k and l all different). For $\overline{\psi_{ij}\psi_{jk}}$ we obtain

$$
\begin{aligned}
\overline{\psi_{ij}\psi_{jk}} &= \frac{1}{N(N-1)(N-2)} \sum_i \sum_j \sum_k \psi_{ij}\psi_{jk} \\
&\qquad\qquad\qquad\qquad {}_{i\neq j, i\neq k, j\neq k} \\
&= \frac{1}{N(N-1)(N-2)} \left\{ \sum_i \sum_j \sum_k \psi_{ij}\psi_{jk} - \sum_j \sum_k \psi_{kj}\psi_{jk} \right\}. \\
&\qquad\qquad\qquad\qquad {}_{i\neq j, j\neq k} \qquad\qquad {}_{j\neq k}
\end{aligned}
\tag{B.22}
$$

It can be readily checked that the ψ_{ij} have the property

$$
\sum_{\substack{i \\ i\neq j}} \psi_{ij} = 0, \qquad \text{for all } j,
\tag{B.23}
$$

Using this relation, the sum over k in the first term within the curly braces in Equation (B.23) can be verified to equal zero. We have

$$
\overline{\psi_{ij}\psi_{jk}} = -\frac{1}{N-2}\,\overline{\psi_{ij}^2}.
\tag{B.24}
$$

Similarly, one can derive the expression

$$
\begin{aligned}
\overline{\psi_{ij}\psi_{kl}} &= -\frac{2}{N-3}\,\overline{\psi_{ij}\psi_{jk}} \\
&= \frac{2}{(N-2)(N-3)}\,\overline{\psi_{ij}^2}.
\end{aligned}
\tag{B.25}
$$

The conditional expected values of the products of random variables (i.e. their averages with respect to the possible choices of D) can be found by straight-forward calculation, involving the relative number of times the indices in the products are within D and D^c. The results are

$$
E_c(A_{12}^2) = \frac{2}{N(N-1)} \times \frac{2(N-1)(N-2)}{N_1 N_2 (N_1-1)(N_2-1)},
\tag{B.26}
$$

$$E_c(A_{12}A_{23}) = -\frac{2}{N(N-1)(N-2)} \times \frac{2(N-1)(N-2)}{N_1 N_2(N_1-1)(N_2-1)}, \quad \text{(B.27)}$$

and

$$E_c(A_{12}A_{34}) = \frac{4}{N(N-1)(N-2)(N-3)} \times \frac{2(N-1)(N-2)}{N_1 N_2(N_1-1)(N_2-1)}. \quad \text{(B.28)}$$

After substitution of these expected values into Equation (B.21) the final expression for the conditional variance is

$$
\begin{aligned}
Var_c\{\widehat{Q}\} &= \left(1 + \frac{2}{(N-2)} + \frac{2}{(N-2)(N-3)}\right) \\
&\quad \times \frac{2(N-1)(N-2)}{N_1 N_2(N_1-1)(N_2-1)} \overline{\psi_{ij}^2} \quad \text{(B.29)} \\
&= \frac{2(N-1)^2(N-2)}{N_1 N_2(N_1-1)(N_2-1)(N-3)} \overline{\psi_{ij}^2}.
\end{aligned}
$$

We can write $\overline{\psi_{ij}^2}$ as

$$\overline{\psi_{ij}^2} = \overline{H_{ij}^2} - 2\frac{N-2}{(N-1)} \overline{g_i^2}, \quad \text{(B.30)}$$

using Equations (B.16)–(B.18).

Appendix C

Spectral Properties of the Solvable Model

This appendix is concerned with some of the details of the derivation of the spectrum of eigenvalues and the integral over the Lyapunov exponents for the solvable model described in Chapter 9.

Appendix C.1. Spectrum of Absolute Values of Eigenvalues

In this section we describe the relation between $\gamma'(s)$, the spectrum of ordered eigenvalues, and $\bar{\gamma}(s)$, the spectrum of ordered absolute values of eigenvalues. We assume, without loss of generality, that $\alpha > 0$ and $0 < \epsilon \leq 1$. To proceed, it is convenient to consider the inverse functions $\gamma'^{-1}(s)$ and $\bar{\gamma}^{-1}(s)$ of $\gamma'(s)$ and $\bar{\gamma}(s)$ respectively. Note that $\gamma'^{-1}(s)$ is the relative number of eigenvalues larger than s. It follows that $1 - \gamma'^{-1}(s)$ is the relative number of eigenvalues smaller than or equal to s. Similarly, $\bar{\gamma}^{-1}(s)$ is the relative number of eigenvalues which, in absolute value, are larger than s. Inverting the function $\gamma'(s)$ in Equation (9.13) leads to

$$\gamma'^{-1}(s) = \pi^{-1} \arccos((s/\alpha - 1)/\epsilon + 1). \qquad (\text{C.1})$$

As long as s is larger than $|\gamma'(1)| = |\alpha(1 - 2\epsilon)|$, the absolute value of the most negative eigenvalue, there are no contributions from negative eigenvalues. The functions $\gamma'^{-1}(s)$ and $\bar{\gamma}^{-1}(s)$ then coincide. If s is smaller than or equal to $|\gamma'(1)|$, then $\bar{\gamma}^{-1}(s)$ has a contribution from positive as well as negative exponents: $\bar{\gamma}^{-1}(s) = \gamma'^{-1}(s) + 1 - \gamma'^{-1}(-s)$.

Keeping track of the situations where these two possibilities occur, still assuming $\alpha > 0$ and $0 < \epsilon \leq 1$, the results can be summarized as

$$\bar{\gamma}^{-1}(s) = \begin{cases} \gamma'^{-1}(s) & \text{if } 0 < \epsilon < \frac{1}{2} \text{ or } \alpha \geq s \geq |\alpha(2\epsilon - 1)| \\ \gamma'^{-1}(s) + 1 - \gamma'^{-1}(-s) & \text{if } \frac{1}{2} \leq \epsilon \leq 1 \text{ and } 0 \leq s < |\alpha(2\epsilon - 1)|. \end{cases} \qquad (\text{C.2})$$

Using the symmetries described by Equations (9.14) and (9.15), an expression for $\bar{\gamma}^{-1}(s)$ can be found in all cases where ϵ and α are both unequal to zero.

Appendix C.2. Integrated Spectrum of Lyapunov Exponents

Next we derive an expression for the integral I over the spectrum of Lyapunov exponents for the solvable model described in Chapter 9. By Equation

(9.22) we have

$$I = \int_0^1 \ln(|p + q\cos\pi s|)\, ds \qquad (C.3)$$

which is preserved under the transformations $p \mapsto -p$ and $q \mapsto -q$. Notice that the integral I does not exist if p and q are both equal to 0.

It is convenient to rewrite Equation (C.3) as

$$I = \frac{1}{2} \int_0^1 \ln(p + q\cos\pi s)^2\, ds. \qquad (C.4)$$

After substitution of $s = 2x/\pi$ we obtain

$$I = \frac{1}{\pi} \int_0^{\frac{1}{2}\pi} \ln(p + q\cos 2x)^2\, dx. \qquad (C.5)$$

Application of the identity $\cos 2\alpha = 1 - 2\sin^2\alpha$ gives

$$I = \frac{1}{\pi} \int_0^{\frac{1}{2}\pi} \ln(p + q - 2q\sin^2 x)^2\, dx \qquad (C.6)$$

which results in

$$I = \begin{cases} \ln|p|, & \text{if } q = 0,\, p \neq 0 \\ \frac{1}{\pi} \int_0^{\frac{1}{2}\pi} \left\{ \ln(2q)^2 + \ln\left(\frac{p+q}{2q} - \sin^2 x\right)^2 \right\} dx, & \text{if } q \neq 0. \end{cases} \qquad (C.7)$$

Using the preservation of I with respect to $q \mapsto -q$ and $p \mapsto -p$ we find

$$I = \begin{cases} \ln|p|, & \text{if } q = 0,\, p \neq 0 \\ \ln 2 + \ln|q| + \frac{1}{\pi} \int_0^{\frac{1}{2}\pi} \ln(\frac{|p|}{2|q|} + \frac{1}{2} - \sin^2 x)^2\, dx, & \text{if } q \neq 0. \end{cases} \qquad (C.8)$$

The integral in this expression can be found using the relation

$$\int_0^{\frac{1}{2}\pi} \ln(\beta^2 - \sin^2 x)^2\, dx = \begin{cases} -2\pi \ln 2, & \text{if } \beta^2 \leq 1 \\ 2\pi \ln \frac{\beta + \sqrt{\beta^2 - 1}}{2}, & \text{if } \beta^2 > 1 \end{cases} \qquad (C.9)$$

(see Gradshteyn and Ryzhik, 1980 (Equation 4.226;1)).

We obtain

$$I = \begin{cases} \ln|p|, & \text{if } q = 0,\, p \neq 0 \\ \ln|q| - \ln 2 + 2\ln\left(\sqrt{\frac{|p|}{2|q|} + \frac{1}{2}} + \sqrt{\frac{|p|}{2|q|} - \frac{1}{2}}\right), & \text{if } |p| > |q|,\, q \neq 0 \\ \ln|q| - \ln 2, & \text{if } |p| \leq |q|,\, q \neq 0. \end{cases} \qquad (C.10)$$

This can be written as

$$
I = \begin{cases} \ln |p|, & \text{if } q = 0,\, p \neq 0 \\ \ln |q| - \ln 2 + \ln \left(\frac{|p|}{|q|} + \sqrt{\frac{|p|^2}{|q|^2} - 1} \right), & \text{if } |p| > |q|,\, q \neq 0 \\ \ln |q| - \ln 2, & \text{if } |p| \leq |q|,\, q \neq 0 \end{cases} \qquad \text{(C.11)}
$$

or

$$
I = \begin{cases} \ln |p|, & \text{if } q = 0,\, p \neq 0 \\ \ln \left(|p| + \sqrt{|p|^2 - |q|^2} \right) - \ln 2, & \text{if } |p| > |q|,\, q \neq 0 \\ \ln |q| - \ln 2, & \text{if } |p| \leq |q|,\, q \neq 0. \end{cases} \qquad \text{(C.12)}
$$

Taking together the first two cases we obtain

$$
I = \begin{cases} \ln \left(|p| + \sqrt{|p|^2 - |q|^2} \right) - \ln 2, & \text{if } |p| > |q| \\ \ln |q| - \ln 2, & \text{if } |p| \leq |q|,\, q \neq 0. \end{cases} \qquad \text{(C.13)}
$$

REFERENCES

Abarbanel, H. D. I. (1993). Nonlinearity and chaos at work. *Nature*, **364**, 672–673.

Abramowitz, M. and Stegun, I. A. (1965). *Handbook of mathematical functions.* Dover Publications, New York.

Aeyels, D. (1981). Generic observability of differentiable systems. *SIAM J. Control and Optimization*, **19**, 595–603.

Aeyels, D. (1982). Global observability of Morse-Smale vectorfields. *J. Diff. Eqns*, **45**, 1–15.

Albano, A. M., Rapp, P. E. and Passamante, A. (1995). Kolmogorov-Smirnov test distinguishes attractors with similar dimensions. *Phys. Rev. E*, **52**, 196–206.

Aliev, R. R. and Rovinsky, A. B. (1992). Spiral waves in the homogeneous and inhomogeneous Belousov-Zhabotinsky reaction. *J. Phys. Chem.*, **96**, 732–736.

Allessie, M. A., Bonke, F. I. and Schopman, F. J. G. (1977). Circus movement in rabbit atrial muscle as a mechanism of tachycardia III. the "leading circle" concept: a new model circus movement in cardiac tissue without the involvement of an anatomical obstacle. *Circ. Res.*, **41**, 9–18.

Anderson, N. H., Hall, P. and Titterington, D. M. (1994). Two-sample test statistics for measuring the discrepancies between two multivariate probability density functions using kernel-based estimates. *J. Multivariate Anal.*, **50**, 41–54.

Babloyantz and Destexhe, A. (1988). Is the normal heart a periodic oscillator? *Biol. Cybern.*, **58**, 203–211.

Badii, R. (1989). Conservation laws and thermodynamic formalism for dissipative dynamical systems. *Riv. Nuovo Cimento*, **12**, 1–72.

Bär, M. and Eiswirth, M. (1993). Turbulence due to spiral breakup in continuous excitable medium. *Phys. Rev. E*, **48**, R1635–R1637.

Barkley, D. (1991). A model for fast computer simulation of waves in excitable media. *Physica D*, **49**, 61–70.

Barkley, D., Kness, M. and Tuckerman, L. S. (1990). Spiral wave dynamics in a simple model of excitable media: Transition from simple to compound rotation. *Phys. Rev.* A, **42**, 2489–2492.

Bauer, M., Heng, H. and Martienssen, W. (1993). Characterization of spatiotemporal chaos from time series. *Phys. Rev. Lett.*, **71**, 521–524.

Bélair, J., Glass, L. and Heiden, U. an der (1995). Dynamical disease: Identification, temporal aspects and treatment strategies of human illness. *Chaos*, **5**, 1–7.

Benedicks, M. (1994). New developments in the ergodic theory of nonlinear dynamical systems. *Phil. Trans. R. Soc. Lond.* A, **346**, 145–157.

Bickel, P. J. (1969). A distribution free version of the Smirnov two sample test in the p-variate case. *Ann. Math. Statist.*, **40**, 1–23.

Biktashev, V. N. and Holden, A. V. (1998). Deterministic Brownian motion in the hypermeander of spiral waves. *Physica* D, **116**.

Billingsley, P. (1979). *Probability and measure*. Wiley, New York.

Box, G. E. P. and Jenkins, G. M. (1976). *Time series analysis*. Holden-Day, San Fransisco.

Brock, W. A., Dechert, W. D. and Scheinkman, J. (1987). A test for independence based on the correlation dimension. Technical report 8702. Social Systems Research Institute, Univ. of Wisconsin, Madison.

Broer, H. W. and Dumortier, F. (1991). *Structures in dynamics*. North-Holland, Amsterdam.

Buzug, T. and Pfister, G. (1992). Comparison of algorithms calculating optimal parameters for delay time coordinates. *Physica* D, **58**, 127.

Caputo, J. G. and Atten, P. (1986). Determination of attractor dimension and entropy for various flows: An experimentalist's viewpoint. In *Dimensions and entropies in chaotic systems* (ed. G. Mayer-Kress), pp. 180–190. Springer, New York.

Caputo, J. G. and Atten, P. (1987). Metric entropy: an experimental means for characterizing and quantifying chaos. *Phys. Rev.* A, **35**, 1311–1316.

Carlstein, E. (1986). The use of subseries values for estimating the variance of a general statistic from a stationary sequence. *Ann. Statist.*, **14**, number 3, 1171–1179.

Casdagli, M., Eubank, S., Farmer, J. D. and Gibson, J. (1991). State space reconstruction in the presence of noise. *Physica* D, **51**, 52–98.

Cessie, S. le and Houwelingen, J. C. van (1993). Building logistic models by means of a non parametric goodness fit test: a case study. *Statist. Neerl.*, **47**, 97–109.

Chan, K.-S. (1996). On the validity of the method of surrogate data. Technical Report No. 248. Fields Institute of Communications.

Chan, K.-S. and Tong, H. (1994). A note on noisy chaos. *J. R. Statist. Soc.* B, **56**, number 2, 301–311.

Chan, K.-S., Tong, H. and Stenseth, N. (1997). Analyzing short time series data from periodically fluctuating rodent populations by threshold models: a nearest block bootstrap approach. Technical Report. University of Iowa.

Cheng, B. and Tong, H. (1992). On consistent non-parametric order determination and chaos. *J. R. Statist. Soc.* B, **54**, 427–449.

Collet, P. and Eckmann, J.-P. (1980). *Iterated Maps on the Interval as Dynamical System*. Birkhäuser, Basel.

Courtemanche, M. and Winfree, A. T. (1991). Re-entrant rotating waves in a Beeler-Reuter based model of two-dimensional cardiac electrical activity. *Int. J. Bifurcation and Chaos*, **1**, 431–444.

Cross, M. C. and Hohenberg, P. C. (1993). Pattern formation outside of equilibrium. *Rev. Mod. Phys.*, **65**, 851–1112.

Cutler, C. D. (1993). A review of the theory and estimation of fractal dimension. In *Dimension estimation and models* (ed. H. Tong), Nonlinear Time Series and Chaos, volume 1. World Scientific, Singapore.

Cvitanović, P. (1989). *Universality in chaos*, Second edn. Adam Hilger, Bristol.

Davidenko, J. M., Kent, P. and Jalife, J. (1991). Spiral waves in normal isolated ventricular muscle. *Physica* D, **49**, 182–197.

Davidenko, J. M., Pertsov, A. V., R. Salomonsz, W. Baxter and Jalife, J. (1992). Stationary and drifting spiral waves of excitation in isolated cardiac muscle. *Nature*, **355**, 349–351.

Davidov, V. A. and Zykov, V. S. (1993). Spiral autowaves in a round excitable media. *JETP*, **76**, 414–419.

Debussche, A. and Marion, M. (1992). On the construction of families of approximate inertial manifolds. *J. Diff. Eqns*, **100**, 173–201.

Denker, M. and Keller, G. (1983). On U-statistics and v. Mises' statistics for weakly dependent processes. *Z. Wahrscheinlichkeitstheorie verw. Gebiete*, **64**, 505–522.

Denker, M. and Keller, G. (1986). Rigorous statistical procedures for data from dynamical systems. *J. Stat. Phys.*, **44**, 67–93.

Diks, C., Houwelingen, J. C. van, Takens, F. and DeGoede, J. (1995). Reversibility as a criterion for discriminating time series. *Phys. Lett. A*, **201**, 221–228.

Diks, C., Zwet, W. R. van, Takens, F. and DeGoede, J. (1996). Detecting differences between delay vector distributions. *Phys. Rev. E*, **53**, 2169–2176.

Dumont, P. S. and Brumer, P. (1988). Characteristics of power spectra for regular and chaotic systems. *J. Chem. Phys.*, **88**, 1481–1496.

Eckmann, J.-P., Kamphorst, S. O. and Ruelle, D. (1987). Recurrence plots of dynamical systems. *Europhys. Lett.*, **4**, 973–977.

Farmer, J. D., Ott, E. and Yorke, J. A. (1983). The dimension of chaotic attractors. *Physica D*, **7**, 153–180.

Feller, W. (1966). *An introduction to probability theory and its applications*, volume 2. Wiley, New York.

Fife, P. C. (1985). Understanding the patterns in the BZ reagent. *J. Stat. Phys.*, **39**, 687–703.

Frank, M., Blank, H.-R., Heindl, J., Kaltenhäuser, M., Köchner, H., , Kreissche, W., Müller, N., Pocher, S., Sporer, R. and Wagner, T. (1993). Improvement of K_2-entropy calculations by means of dimension scaled distances. *Physica D*, **65**, 359–364.

Fraser, A. M. and Swinney, H. L. (1986). Independent coordinates for strange attractors from mutual information. *Phys. Rev. A*, **33**, 1134–1140.

Gaspard, P. and Wang, X.-J. (1993). Noise, chaos and (ϵ, τ)-entropy per unit time. *Physics Reports*, **235**, 291–343.

Ghez, J. M., Orlandini, E., Tesi, M.-C. and Vaienti, S. (1993). Dynamical integral transform on fractal sets and the computation of entropy. *Physica D*, **63**, 282–298.

Ghez, J. M. and Vaienti, S. (1992a). Integrated wavelets on fractal sets: I. the correlation dimension. *Nonlinearity*, **5**, 777–790.

Ghez, J. M. and Vaienti, S. (1992b). Integrated wavelets on fractal sets: II. the generalized dimensions. *Nonlinearity*, **5**, 791–804.

Giannakis, G. B. and Tsatsanis, M. K. (1994). Time-domain tests for Gaussianity and time-reversibility. *IEEE Trans. on Signal Processing*, **42**, 3460–3472.

Gradshteyn, I. S. and Ryzhik, I. M. (1980). *Table of integrals, series, and products*, corrected and enlarged edn. Academic Press, New York.

Grassberger, P. (1983). On the fractal dimension of the Hénon attractor. *Phys. Lett.* A, **97**, 224–226.

Grassberger, P. (1988). Finite sample corrections to entropy and dimension estimates. *Phys. Lett.* A, **128**, 369–373.

Grassberger, P. (1990). An optimized box-assisted algorithm for fractal dimensions. *Phys. Lett.* A, **148**, 63–68.

Grassberger, P., Hegger, R., Kantz, H., Schaffrath, C. and Schreiber, Th. (1993). On noise reduction methods for chaotic data. *Chaos*, **3**, 127–140.

Grassberger, P. and Procaccia, I. (1983a). Characterization of strange attractors. *Phys. Rev. Lett.*, **50**, 346–349.

Grassberger, P. and Procaccia, I. (1983b). Measuring the strangeness of strange attractors. *Physica* D, **9**, 189–208; reprinted in Hao (1990).

Grassberger, P., Schreiber, T. and Schaffrath, C. (1991). Nonlinear time sequence analysis. *Int. J. Bifurcation and Chaos*, **1**, 521–547.

Grigoriev, R. O. (1998). Solvable model for spatiotemporal chaos. *Physical Review* E, **57**, 388–396.

Hall, P., Horowitz, J. L. and Jing, B.-Y. (1995). On blocking rules for the bootstrap with dependent data. *Biometrika*, **82**, number 3, 561–574.

Halsey, T. C., Jensen, M. H., Kadanoff, L. P., Procaccia, I. and Shraiman, B. I. (1986). Fractal measures and their singularities: The characterization of strange sets. *Phys. Rev.* A, **33**, 1141–1151; reprinted in Cvitanović (1989); Hao (1990).

Hao, B.-L. (1984). *Chaos*. World Scientific, Singapore.

Hao, B.-L. (1990). *Chaos II*. World Scientific, Singapore.

Heijden, M. J. van der, Diks, C., Pijn, J. P. M. and Velis, D. N. (1996). Time reversibility of intracerebral human EEG recordings in mesial temporal lobe epilepsy. *Physics Letters* A, **216**, 283–288.

Hénon, M. (1976). A two-dimensional mapping with a strange attractor. *Commun. Math. Phys.*, **50**, 69–77; reprinted in Cvitanović (1989).

Hentschel, H. G. E. and Procaccia, I. (1983). The infinite number of generalized dimensions of fractals and strange attractors. *Physica* D, **8**, 435–444; reprinted in Hao (1990).

Herzel, H., Ebeling, W. and Schulmeister, T. (1987). Nonuniform chaotic dynamics and effects of noise in biochemical systems. *Z. Naturforsch.* A, **42**, 136–142.

Herzel, H., Schmitt, A. O. and Ebeling, W. (1994). Finite sample effects in sequence analysis. *Chaos Solitons and Fractals*, **4**, 97–113.

Hoekstra, B. P. T., Diks, C. G. H., , Allessie, M. A. and DeGoede, J. (1997). Nonlinear analysis of the pharmacological conversion of sustained atrial fibrillation in conscious goats by the class Ic drug cibenzoline. *Chaos*, **7**, 430–446.

Holzfuss, J. and Kadtke, J. (1993). Global nonlinear noise reduction using radial basis functions. *Int. J. Bifurcation and Chaos*, **3**, number 3, 589–595.

Hunt, B. R., Sauer, T. and Yorke, J. A. (1992). Prevalence: a translation-invariant "almost every" on infinite-dimensional spaces. *Bull. Am. Math. Soc.*, **27**, 217–238.

Isliker, H. (1992). A scaling test for correlation dimensions. *Phys. Lett.* A, **169**, 313–322.

Jackson, E. A. and Kodogeorgiou, A. (1992). A coupled Lorenz-cell model of Rayleigh-Bénard turbulence. *Physics Letters* A, **168**, 270–275.

Jahnke, W. and Winfree, A. T. (1991). A survey of spiral-wave behaviors in the oregonator model. *Int. J. Bifurcation and Chaos*, **1**, 445–466.

Jensen, J. L. (1993). Chaotic dynamical systems with a view towards statistics: a review. In *Networks and chaos – statistical and probabilistic aspects* (eds O. E. Barndorff-Nielsen, J. L. Jensen and W. S. Kendall), pp. 201–250. Chapman & Hall, London.

Kampen, N. G. van (1992). *Stochastic processes in physics and chemistry*, second edn. North-Holland, Amsterdam.

Kaneko, K. (1985a). Kinks and turbulence in coupled map lattices. In *Dynamical problems in soliton systems* (ed. S. Taneko). Springer Verlag, Berlin.

Kaneko, K. (1985b). Spatiotemporal intermittency in coupled map lattices. *Prog. Theor. Phys.*, **74**, 1033.

Kaneko, K. (1992). Overview of coupled map lattices. *Chaos*, **2**, 279–282.

Kantz, H. (1994). Quantifying the closeness of fractal measures. *Phys. Rev. E*, **49**, 5091–5097.

Kantz, H. and Schreiber, Th. (1997). *Nonlinear time series analysis*. Cambridge nonlinear science series 7. Cambridge University Press, Cambridge.

Kapitaniak, T. (1990). *Chaos in systems with noise*, second edn. World Scientific, Singapore.

Kaplan, D. and Glass, L. (1995). *Understanding nonlinear dynamics*. Springer Verlag, Berlin.

Kaplan, D. T. and Glass, L. (1992). Direct test for determinism in a time series. *Phys. Rev. Lett.*, **68**, 427–430.

Kaplan, D. T. and Glass, L. (1993). Coarse-grained embeddings of time series: Random walks, Gaussian random processes, and deterministic chaos. *Physica D*, **64**, 431–454.

Kaplan, J. L. and Yorke, J. A. (1979). Chaotic behavior of multidimensional difference equations. In *Functional Differential Equations and Approximation of Fixed Points* (eds H. O. Walter and H.-O. Peitgen), volume 730, pp. 204–227. Springer Verlag, Berlin.

Karma, A. (1990). Meandering transition in two-dimensional excitable media. *Phys. Rev. Lett.*, **65**, 2824–2827.

Karma, A. (1993). Spiral breakup in model equations of action potential propagation in cardiac tissue. *Phys. Rev. Lett.*, **71**, 1103–1106.

Keener, J. P. and Tyson, J. J. (1986). Spiral waves in the Belousov-Zhabotinskii reaction. *Physica D*, **21**, 307–324.

Keller, G., Künzle, M. and Nowicki, T. (1992). Some phase transitions in coupled map lattices. *Physica D*, **59**, 39–51.

Kennel, M. B. and Isabelle, S. (1992). Method to distinguish chaos from colored noise and to determine embedding parameters. *Phys. Rev.* A, **46**, 3111–3118.

Kifer, Y. (1986). *Ergodic theory of random transformations.* Birkhäuser, Boston.

Kloeden, P. E. and Platen, E. (1995). *Numerical solution of stochastic differential equations.* Applications of Mathematics 23, 2nd edn. Springer, Berlin.

Konings, K. T. S., Kirchoff, C. J. H. J., Smeets, J. R. L. M., Wellens, H. J. J. and Allessie, M. A. (1994). High-density mapping of electrically induced atrial fibrillation in humans. *Circulation*, **89**, 1665–1680.

Kostelich, E. J. and Schreiber, Th. (1993). Noise reduction in chaotic time-series data: A survey of common methods. *Phys. Rev.* E, **48**, 1752–1762.

Kostelich, E. J. and Yorke, J. A. (1990). Noise reduction: Finding the simplest dynamical system consistent with data. *Physica* D, **41**, 183–196.

Kreiss, J.-P., Neumann, M. H. and Yao, Q. (1998). Bootstrap tests for simple structure in nonparametric time series regression. Technical Report UKC/IMS/98/10. Institute of Mathematics and Statistics, University of Kent, Canterbury, U. K.

Künsch, H. R. (1989). The jackknife and the bootstrap for general stationary observations. *Ann. Statist.*, **17**, 1217–1241.

Lall, U. and Sharma, A. (1996). A nearest neighbor bootstrap for time series resampling. *Water Resources Research*, **32**, 679–693.

Lasota, A. and Mackay, M. C. (1994). *Chaos, fractals and noise.* Applied Mathematical Sciences, 2nd edn, volume 97. Springer, New York.

Lawrance, A. J. (1991). Directionality and reversibility in time series. *Int. Statist. Rev.*, **59**, 67–79.

Lawrance, A. J. and Spencer, N. M. (1995). Statistical aspects of curved chaotic map models and their stochastic reversals. Technical Report. School of Mathematics and Statistics, University of Birmingham, U. K.

Ljung, L. (1987). *System Identification: Theory for the User.* Prentice-Hall, Englewood Cliffs.

Lorenz, E. N. (1963). Deterministic nonperiodic flow. *J. Atmos. Sci.*, **20**, 130–141; reprinted in Cvitanović (1989).

Lugosi, E. (1989). Analysis of meandering in Zykov kinetics. *Physica D*, **40**, 331–337.

Manneville, P. (1990). *Dissipative structures in weak turbulence.* Academic Press, New York.

Mayer-Kress, G. and Kaneko, K. (1989). Spatiotemporal chaos and noise. *J. Stat. Phys.*, **54**, 1489–1508.

Mees, A. (1993). Parsimonious dynamical reconstruction. *Int. J. Bifurcation and Chaos*, **3**, number 3, 669–675.

Melo, W. de and Strien, S. van (1993). *One-dimensional dynamics.* Ergebnisse der Mathematik und ihrer Grenzgebiete, volume 3. Springer-Verlag, Berlin.

Meron, E. (1992). Pattern formation in excitable media. *Physics Reports*, **218**, 1–66.

Moe, G. K., Rheinboldt, W. C. and Abildskov, J. A. (1964). A computer model of atrial fibrillation. *Am. Heart J.*, **67**, 200–220.

Murray, J. D. (1989). *Mathematical biology.* Springer, New York.

Nicolis, C., Nicolis, G. and Wang, Q. (1992). Sensitivity to initial conditions in spatially extended systems. *Int. J. Bifurcation and Chaos*, **2**, 263–269.

Noakes, L. (1991). The Takens embedding theorem. *Int. J. Bifurcation and Chaos*, **1**, 867–872.

Nychka, D., Ellner, S., Gallant, A. R. and McCaffrey, D. (1992). Finding chaos in noisy systems. *J. R. Statist. Soc.* B, **54**, 399–426.

Olofsen, E., DeGoede, J. and Heijungs, R. (1992). A maximum likelihood approach to correlation dimension and entropy estimation. *Bull. Math. Biol.*, **54**, 45–58.

Oltmans, H. and Verheijen, P. J. T. (1997). Influence of noise on power-law scaling functions and an algorithm for dimension estimations. *Phys. Rev. E*, **56**, 1160–1170.

Ott, E. (1993). *Chaos in Dynamical Systems.* Cambridge University Press, Cambridge.

Ott, E. and Hanson, J. D. (1981). The effect of noise on the structure of strange attractors. *Phys. Lett.* A, **85**, 20.

Ott, E., Sauer, T. and Yorke, J. A. (1994). *Coping with Chaos*. Wiley, New York.

Ott, E., Withers, W. D. and Yorke, J. A. (1984). Is the dimension of chaotic attractors invariant under coordinate change? *J. Stat. Phys.*, **36**, 687–697.

Ott, E., Yorke, E. D. and Yorke, J. A. (1985). A scaling law. How an attractor's volume depends on noise level. *Physica* D, **16**, 62–78.

Packard, N. H., Crutchfield, J. P., Farmer, J. D. and Shaw, R. S. (1980). Geometry from a time series. *Phys. Rev. Lett.*, **45**, 712–716.

Paladin, G. and Vulpiani, A. (1987). Anomalous scaling laws in multifractal objects. *Phys. Rep.*, **156**, 147–225.

Panfilov, A. and Hogeweg, P. (1993). Spiral breakup in a modified FitzHugh-Nagumo model. *Phys. Lett.* A, **176**, 295–299.

Panfilov, A. V. and Holden, A. V. (1991). Spatiotemporal irregularity in a two-dimensional model of cardiac tissue. *Int. J. Bifurcation and Chaos*, **1**, 219–225.

Parker, T. S. and Chua, L. O. (1989). *Practical numerical algorithms for chaotic systems*. Springer, Berlin.

Peinke, J., Parisi, J., Rössler, O. E. and Stoop, R. (1992). *Encounter with chaos; Self-organized hierarchical complexity in semiconductor experiments*. Springer Verlag, New York.

Pesin, Ya. B. (1977). Characteristic lyapunov exponents and smooth ergodic theory. *Russ. Math. Surv.*, **32**, 55–104.

Petersen, K. (1983). *Ergodic theory*. Cambridge University Press, Cambridge.

Pezard, L., Martineri, J., Müller-Gerking, J., Varela, F. J. and Renault, B. (1996). Entropy quantification of human brain spatio-temporal dynamics. *Physica* D, **96**, 344.

Plesser, T. and Müller, K.-H. (1995). Fourier analysis of the complex motion of spiral tips in excitable media. *Int. J. Bifurcation and Chaos*, **5**, 1071–1084.

Press, W. H., Teukolsky, S. T., Vetterling, W. T. and Flannery, B. P. (1992). *Numerical Recipes in C: the art of scientific computing,* 2nd edn. Cambridge University Press, Cambridge.

Price, C. P. and Prichard, D. (1993). On the embedding statistic. *Phys. Lett. A,* **184,** 83–87.

Priestley, M. B. (1981). *Spectral analysis and time series,* vols I and II. Academic Press, London.

Pukelsheim, F. (1994). The three sigma rule. *Am. Statist.,* **48,** 88–91.

Radu, B. (1993). Attractors with $D_c \neq D_c(1)$ and $D_c \neq D_H$. *Physica* D, **68,** 281–282.

Rensma, P. L., Allessie, M. A., Lammers, W. J. E. P., Bonke, F. I. M. and Schalij, M. J. (1988). Length of excitation wave and susceptibility to reentrant atrial arrhythmias in normal conscious dogs. *Circ. Res.,* **62,** 395–410.

Renyi, A. (1971). *Probability theory.* North-Holland, Amsterdam.

Roberts, J. A. G. and Quispel, G. R. W. (1992). Chaos and time-reversal symmetry. Order and chaos in reversible dynamical systems. *Physics Reports,* **216,** 64–177.

Robinson, P. M. and Velasco, C. (1996). Autocorrelation-robust inference. *Econometrics,* **316.**

Romano, J. P. (1988). A bootstrap revival of some nonparametric distance tests. *J. Am. Statist. Assoc.,* **83,** 698–708.

Rössler, O. E. (1976). An equation for continuous chaos. *Phys. Lett.* A, **57,** 397–398.

Ruelle, D. (1980). Strange attractors. *Math. Intelligencer,* **2,** 126–137.

Ruelle, D. and Takens, F. (1971). On the nature of turbulence. *Commun. Math. Phys.,* **20,** 167–192.

Russel, D. A., Hanson, J. D. and Ott, E. (1980). Dimension of strange attractors. *Phys. Rev. Lett.,* **45,** 1175–1178.

Sauer, T. (1992). A noise reduction method for signals from nonlinear systems. *Physica* D, **58,** 193–201.

Sauer, T. and Yorke, J. A. (1993). How many delay coordinates do you need? *Int. J. Bifurcation and Chaos*, **3**, 737–744.

Sauer, T., Yorke, J. A. and Casdagli, M. (1991). Embedology. *J. Stat. Phys.*, **65**, 579–616.

Schouten, J. C., Takens, F. and Bleek, C. M. van den (1994a). Estimation of the dimension of a noisy attractor. *Phys. Rev A*, **50**, 1851–1861.

Schouten, J. C., Takens, F. and Bleek, C. M. van den (1994b). Maximum-likelihood estimation of the entropy of an attractor. *Phys. Rev. E*, **49**, 126–129.

Schreiber, T. (1997). Influence of Gaussian noise on the correlation exponent. *Phys. Rev. E*, **56**, 274–277.

Schreiber, Th. (1993a). Determination of the noise level of chaotic time series. *Phys. Rev. E*, **48**, R13–R16.

Schreiber, Th. (1993b). Extremely simple nonlinear noise-reduction method. *Phys. Rev. E*, **4**, 2401–2404.

Schuster, H. G. (1988). *Deterministic chaos: An introduction.* VCH, Weinheim.

Serfling, R. J. (1980). *Approximation theorems of mathematical statistics.* Wiley, New York.

Smith, L. A. (1992a). Identification and prediction of low dimensional dynamics. *Physica D*, **58**, 50–76.

Smith, R. L. (1992b). Estimating dimension in noisy chaotic time series. *J. R. Statist. Soc. B*, **54**, 329–351.

Solé, R. V., Bascompte, J. and Valls, J. (1992). Nonequilibrium dynamics in lattice ecosystems: chaotic stability and dissipative structures. *Chaos*, **2**, 387–395.

Stockis, J.-P. and Tong, H. (1998). On the statistical inference of a machine-generated AR(1) model. *J. R. Statist. Soc. B*, **60**, 781–796.

Takens, F. (1981). Detecting strange attractors in turbulence. In *Dynamical Systems and Turbulence, Warwick 1980 (Lecture Notes in Mathematics)* (eds D. A. Rand and L.-S. Young), volume 898, pp. 366–381. Springer, Berlin.

Takens, F. (1985). On the numerical determination of the dimension of an attractor. In *Dynamical systems and bifurcations, Groningen 1984 (Lecture Notes in Mathematics)* (eds Braaksma, B. L. J., Broer, H. W. and F. Takens), volume 1125, pp. 99–106. Springer, Berlin.

Takens, F. (1993). Detecting nonlinearities in stationary time series. *Int. J. Bifurcation and Chaos*, **3**, 241–256.

Takens, F. (1996). The effect of small noise on systems with chaotic dynamics. In *Stochastic and spatial structures of dynamical systems* (eds S. J. van Strien and S. M. Verduyn Lunel), Verhandelingen KNAW, Afd. Natuurkunde, volume 45, pp. 3–15. North-Holland, Amsterdam.

Theiler, J. (1986). Spurious dimensions from correlation algorithms applied to limited time-series data. *Phys. Rev.* A, **34**, 2427–2432.

Theiler, J. (1990). Statistical precision of dimension estimators. *Phys. Rev.* A, **41**, 3038–3051.

Theiler, J., Eubank, S., Longtin, A., Galdrikian, B. and Farmer, J. D. (1992a). Testing for nonlinearity in time series: the method of surrogate data. *Physica D*, **58**, 77–94.

Theiler, J., Galdrikian, B., Longtin, A., Eubank, S. and Farmer, J. D. (1992b). Using surrogate data to detect nonlinearity in time series. In *Nonlinear modeling and forecasting. Proceedings of the workshop held September, 1990, in Santa Fe, New Mexico* (eds M. Casdagli and S. Eubank), pp. 163–188. Addison Wesley, Reading Mass.

Theiler, J. and Lookman, T. (1993). Statistical error in a chord estimator of correlation dimension: The "rule of five". *Int. J. Bifurcation and Chaos*, **3**, 765–771.

Timmer, J. (1995). On surrogate data testing for linearity based on the periodogram. Technical Report. Faculty of Physics, University Freiburg, Germany. Preprint comp-gas/9509003 on http://xyz.lanl.gov.

Tong, H. (1983). Threshold models in non-linear time series analysis. In *Lecture Notes in Statistics*, volume 21. Springer, Berlin.

Tong, H. (1990). *Nonlinear time series analysis: A dynamical system approach.* Oxford University Press, Oxford.

Tong, H. (1992). Some comments on a bridge between nonlinear dynamicists and statisticians. *Physica D*, **58**, 299–303.

Tong, H. (1995). A personal overview of non-linear time series analysis from a chaos perspective. *Scand. J. Statist.*, **22**, 399–445.

Tong, H. and Cheng, B. (1992). A note on one-dimensional chaotic maps under time reversal. *Adv. Appl. Probab.*, **24**, 219–220.

Torcini, A., Politi, A., Puccioni, G. P. and D'Alessandro, G. (1991). Fractal dimensions of spatially extended systems. *Physica* D, **53**, 85–101.

Tsay, R. S. (1986). Nonlinearity tests for time series. *Biometrika*, **73**, 461–466.

Tsimring, L. S. (1993). Nested strange attractors in spatiotemporal chaotic systems. *Phys. Rev.* E, **48**, 3421–3426.

Tyson, J. J., Alexander, K. A., Manoranjan, V. S. and Murray, J. D. (1989). Spiral waves of cyclic AMP in a model of slime mold aggregation. *Physica* D, **34**, 193–207.

Veneziano, D., Moglen, G. E. and Bras, R. L. (1995). Multifractal analysis: Pitfalls of standard procedures and alternatives. *Phys. Rev.* E, **52**, 1387–1398.

Wayland, R., Bromley, D., Pickett, D. and Passamante, A. (1993). Recognizing determinism in a time series. *Phys. Rev. Lett.*, **70**, 580–582.

Weiss, G. (1975). Time-reversibility of linear stochastic processes. *J. Appl. Probab.*, **12**, 831–836.

Wiggins, S. (1990). *Introduction to applied dynamical systems and chaos.* Springer, New York.

Wolf, A., Swift, J. B., Swinney, H. L. and Vastano, J. A. (1985). Determining Lyapunov exponents from a time series. *Physica* D, **16**, 285–317.

Wolfram, S. (1984). Universality and complexity in cellular automata. *Physica* D, **10**, 1–35.

Wright, J. and Schult, R. L. (1993). Recognition and classification of nonlinear chaotic signals. *Chaos*, **3**, 295–304.

Yanagita, T. and Kaneko, K. (1993). Coupled map lattice model for boiling. *Physics Letters* A, **175**, 415–420.

Yao, Q. and Tong, H. (1994a). On prediction and chaos in stochastic systems. *Philos. Trans. Roy. Soc.* A, **348**, number 1688, 357–369.

Yao, Q. and Tong, H. (1994b). Quantifying the influence of initial values on non-linear predictions. *J. R. Statist. Soc.* B, **56,** number 4, 701–725.

Yao, Q. and Tong, H. (1998). A bootstrap detection for operational determinism. *Physica* D, **115,** 49–55.

Yule, G. U. (1927). On a method of investigating periodicities in disturbed series with special reference to Wolfer's sunspot numbers. *Philos. Trans. Roy. Soc. London* A, **226,** 267–298.

Zwet, W. R. van (1984). A Berry-Esseen bound for symmetric statistics. *Z. Wahrscheinlichkeitstheorie Verw. Gebiete*, **66,** 425–440.

Index

Anosov diffeomorphism, 176
AR model, 34
    chaos driven, 38
ARMA model, 34
atrial fibrillation, 23, 44, 107
attractor, 11
    Lorenz, 14
    reconstruction, 16
    Rössler, 15
    strange, 12

bandwidth, 75, 88
Bernoulli map, 173
bilinear form, 56
block
    bootstrap, 38, 39, 75
    method, 65, 86
    randomization, 75, 92
bootstrap, 38, 39, 75
box-counting dimension, 26

capacity dimension, 26
change point, 47
chaos
    analysis, 2, 4, 7, 147
    spatio-temporal, 39, 147, 149,
        162, 173
chaotic, 8, 51
correlation dimension, 18, 97
    coarse-grained, 111, 123
correlation entropy, 18, 50, 97
    coarse-grained, 111, 124
coupled map lattice, 40, 150

delay vectors, 17
dependence, 84
    block method, 65
    covariance, 72

dimension
    box-counting, 26
    capacity, 26
    density, 157
    Kaplan-Yorke, 151
dissipative, 11
doubling map, 10
dynamical invariants, 97
dynamical system, 7
    definition, 9
    discrete/continuous time, 10
    reversible, 175
dynamics
    dissipative, 13
    local, 129
    reconstruction, 16

ElectroEncephaloGram (EEG), 47
evolution, 10
    operator, 9
excitable media, 4, 42, 128
excitation variable, 129

fixed point, 8
    attractor, 179
    unstable, 13
flow, 10
fluidized bed, 108, 166
fractal, 118
    dimension, 1, 22, 26
    measure, 50
    multi-, 26

G-orbit, 58
GARCH model, 71
Gaussian kernel, 99, 125

Hénon map, 11
hyper-meander, 44, 128

hyperbolic, 33

information dimension, 26
    density, 157
information entropy, 26
    density, 160
invariant measure, 17, 33, 167, 176

Kaplan-Yorke dimension, 151
kernel function, 37, 60, 86, 98, 119
    Gaussian, 99, 125
kernel regression, 39

Lebesgue measure, 12, 162
linear Gaussian process, 53
local dynamics, 129
logistic map, 179
Lorenz model, 12
Lyapunov exponent, 27
    stochastic, 36
Lyapunov spectra, 151, 187

map
    Anosov, 176
    Bernoulli, 173
    doubling, 10
    Hénon, 11
    logistic, 7, 179
    reversing, 175
    weakly reversible, 175
meander, 44, 128
    hyper-, 44, 128
    model, 140
measure
    fractal, 50
    invariant, 17, 33, 167, 176
    Lebesgue, 12, 162
    physical, 168
    prediction, 49, 50
    reconstruction, 48, 50, 53, 55,
        81, 84

measurement function, 10, 177
Monte Carlo test, 92
moving average (MA) model, 34
multi-fractal, 26
mutual information, 23

noise
    bounded, 98
    dynamical, 29
    explosion, 33
    external, 32
    Gaussian, 98
    observational, 29, 97, 117
    reduction, 31, 98
    system, 32
nonlinear time series, 28, 49
nonparametric
    modeling, 39
    regression, 39
    test, 82

parametric bootstrap, 51
phase portrait, 178
phase randomization, 51
phase space, 9
physical measure, 168
piece-wise linear model, 153
prediction measure, 49, 50

Rössler model, 15
randomization test, 75, 92
reaction-diffusion model, 128, 147
read-out function, 10
reconstruction
    attractor, 16
    dynamics, 16
    measure, 48, 50, 53, 55, 81, 84
    theorem, 15
    vectors, 17, 182
reconstruction measure, 178
recovery variable, 129, 147

respiration, 108
reversibility, 52, 175

scaling region, 21, 98, 117
sensitive dependence, 11, 14, 36, 46,
        145
shadowing property, 33
solvable model, 149
spatio-temporal chaos, 39, 147, 149,
        162, 173
spiral wave, 128
    break-up, 44, 146
    tip, 129, 130
state space, 9
stationarity, 47, 81
sufficient statistic, 58
surrogate data, 51

test for equivalence of reconstruc-
        tion measures, 81
test for reversibility, 51
Theiler correction, 61, 86, 103, 131
threshold AR (TAR) model, 35
time series
    nonlinear, 28, 49
    reversible, 175
    stationary, 47
tip trajectory, 140

U-statistic, 37, 57, 59

weakly reversible, 175